Bittersweet Brexit

Bittersweet Brexit

The Future of Food, Farming, Land and Labour

Charlie Clutterbuck PhD

First published 2017 by Pluto Press
345 Archway Road, London N6 5AA

www.plutobooks.com

British Library Cataloguing in Publication Data
A catalogue record for this book is available from the British Library

ISBN 978 0 7453 3771 5 Hardback
ISBN 978 0 7453 3770 8 Paperback
ISBN 978 1 7868 0207 1 PDF eBook
ISBN 978 1 7868 0209 5 Kindle eBook
ISBN 978 1 7868 0208 8 EPUB eBook

This book is printed on paper suitable for recycling and made from fully managed and sustained forest sources. Logging, pulping and manufacturing processes are expected to conform to the environmental standards of the country of origin.

Typeset by Stanford DTP Services, Northampton, England

Simultaneously printed in the United Kingdom and United States of America

Contents

List of Photographs, Figures and Tables

PHOTOGRAPHS

FIGURES

TABLES

Foreword

When Charlie suggested I write the foreword to his book on food post Brexit, my immediate reaction was, why me? I'm not an expert on any aspect of the food chain, have no qualification in nutrition or food science and possess only a layperson's knowledge of the politics of food and why we are where we are.

But I guess to some extent that's the point. I'm a working mother with grave concerns about the future of my child and the children I will never ever meet.

I'm a founder of a grassroots movement, Incredible Edible, whose purpose is to get people thinking of the world they live in and what they can do to build something kinder. 'If you eat you're in' is our motto, and we use food to motivate people to challenge a status quo that isn't working for many people across the globe.

So of course Charlie's book is important to people like me who have a stake in how our food is produced, where our food is produced and what impact all that has on the planet we call home.

Over the years Charlie and I have spoken many times of the idiocy of flying beans half way across the planet and the importance of creating sticky-money food economies where the profit stays where the money's spent and you can see how it's produced if you had a mind to.

We've ranted at the idiocy of not using our schools to teach our children about the importance of living soils and respect for all life forms on which human beings depend for their existence.

For me, better use of our public realm and our markets to reconnect people to growing food and seasonality is the first step in rethinking what we want from the food sector and challenging the powers that be to create a system that invests in the wellbeing of all, not only those who can afford the artisan food we find in many upmarket supermarkets.

I'm an optimist and I believe that an informed public can shift the patterns of their life if they are given alternatives that allow them to imagine a better, happier, healthier future for their families.

That's why this book is so important, not just because it informs, but because it challenges us all to take the opportunity Brexit presents to rethink our food systems, rethink our investment in food production and reconnect, locally, with the opportunities to re-skill and re-plan land use, so we are not flying all those beans from one side of the world to the other.

Bring it on.

Pam Warhurst
Founder of Incredible Edible in Todmorden, Yorkshire, Incredible Edible Network – 'If you eat, you're in'

Photo 1 The then Mayor of Todmorden Jayne Booth listens in to Pam Warhurst at the opening of the Aquagarden in the small Yorkshire town.

Acknowledgements

Thanks go to Mark Metcalf for making this book happen, my wife Frances for helping and supporting me, Richard Hooper for the cartoons, Geoff Tansey and Anne Beech for editing, Nick Hayes for helping with the money tree, Steve Leniec for helpful comments, Incredible Edible in Todmorden for inspiration, Sustainable Food friends in the North West, Tim Lang for lots over many years, and Jenny Shepherd, Carol Marshall and Nancy Thompson for proofing and commenting, and thanks to Unite the Union, in particular Jim Mowatt, Director of Education, for backing this book.

Introduction

My book spells out the big changes to come. It draws on my over 40-year experience as an agricultural science worker and trade unionist. I trained as a scientist, gaining three agricultural degrees, including a doctorate from Wye College, then probably the best agricultural college in the world, but since closed down. I learnt about many agricultural science disciplines, from pests and diseases, biochemistry, pesticides, agronomy, horticulture, tropical crops, through to soil science and, in particular, soil zoology.

I also have experience in all parts of the food and farming chain; from working in kitchens and fields, plucking turkeys, picking hops, helping set up the first 'politics of food' group, living on a farm, advising retailers about responsible food, building 'food ethics' into online food supply chains, to setting up a module in a new food entrepreneurial degree on sustainable food.

I have been involved with the farmworkers' union for over 40 years, first writing articles on pesticides for the National Union of Agricultural and Allied Workers in the 1970s, then representing them on the Health and Safety Executive's (HSE) Chemicals in Agriculture (CHEMAG) Committee and then the Agricultural Industry Advisory Committee. I was the North West representative on the National Committee for Rural and Agricultural Sector of Unite, until it merged with the Food Sector of the union.

In Part I, Chapter 1 examines where we are now and how we got here. The main thing I learnt while studying for those degrees in agricultural sciences is that overpopulation is not the problem. Overproduction is. In Chapter 2, I spell out how we may come out of the EU in terms of Brexit itself, with regard to the Single Market and Customs Union. There are more directives applying to the government department responsible for food and farming (DEFRA) than any other department. In Chapter 3, whatever deal is concluded, I discuss how we can still decide whether we are 'Going Global' or 'Buying British'.

Part II looks across a range of food-related issues affecting society. The first of which is trade, the challenges around which I discuss in Chapter

4. Once we step outside the Customs Union, there are 2,000 agricultural products with tariffs that protect farms and food-producers in the EU. If we jump off the EU Tariff Cliff we suddenly face a double whammy – e.g. there may be no tariffs blocking Australian lamb coming to the UK, but tariffs will block our lamb – and other meat – going into the EU. When we have sorted those out, there are 15,000 Processed Agricultural Products (PAPs) each with their own tariffs. When people say we can trade freely under WTO rules, they are kidding everybody.

Throughout the food system we need people to produce and serve the food we eat. In Chapter 5, I look at the conditions under which people labour. Here I make a unique proposal to change the way the £3bn plus EU CAP funds are distributed – we should pay the money to workers on the land, not the owners of the land. At present, over 90 per cent of the EU CAP funds go to those owning more than 10 acres – to do nothing. The more land they own, the more they get – for doing nothing.

How we as a society treat land is crucial for our food and well-being and there is no level ploughing field. As I discuss in Chapter 6, the Countryside Survey, ten years ago, clearly pointed to poor soil management. There are significant losses of carbon from arable soils, and 2 million tonnes of soil lost through erosion each year. I have a particular interest as a soil zoologist and take you on a unique journey among the soil animals.

This leads to Part III of the book, which looks at science. As I call myself an agricultural science worker, I apply my scientific expertise and look at what may happen to four main areas of farm and food science. Chapter 7 recognises that any future food and farm system has to be better for the planet. The EU made sustainable agriculture the number one priority but all the signs are that this will be ignored here after Brexit. My next chapter investigates the biggest cost from the food system – obesity – where I use my experience of developing the World Health Organisation's learning materials for Africa and Asia to combat 'Globesity'. In Chapter 9 my focus is on pesticides. Pesticide controls will become a 'devolved' issue. I represented farmworkers on the HSE for 30 years and was on the government's Advisory Committee on Pesticides (ACP) for 5 years. I choose a couple of controversial chemicals – neonics and glyphosate – to stimulate a debate about what we could do. Chapter 10 deals with the contentious issue of GMOs, as they will figure prominently on the agenda post Brexit, as a newly devolved matter.

Part IV of the book looks to the future. Chapter 11 looks at many of our favourite foods and what may happen to them. The final chapter takes a more general overview, building on the main thrust of the book, which is that we should produce more of the food we eat, rather than dashing round the world 'making deals' when the biggest deal we could make is right under our noses. We produce only just over half of the food we eat. We import $66bn and export $33bn, so our food trade deficit (FTD) is $33bn. If we reduced even part of that import bill by eating food we produce ourselves, that money could be put to good use elsewhere.

I use the words 'we' or 'ours' very loosely to mean nationhood, without being too specific which nation. This is not an academic text, leaving you to google a few words of many of the stories to find more. There are web links to specific issues in the notes, all of which were working in August 2017. When you see $ it means dollars – as that is what international import/export statistics use. Sometimes euros and pounds are quoted, when I can only get the relevant figures in those currencies. I use 'Britain' and the 'United Kingdom' interchangeably, although I know the United Kingdom is 'Britain and Northern Ireland'. I use the term 'global warming' rather than the more popular 'climate change', as it is harder to deny that the globe is warming. Climate changes are unpredictable, whereas the globe is warming very predictably and inexorably.

The main aim of this book is to stimulate a debate about the sort of food and farming we want for our future. While providing a guide to the main issues following Brexit, I make clear where I stand. I expect it to be controversial, so it stimulates dialogue. While using the word 'dialogue', I think 'dialectic' is better, as it recognises how opposing forces can create new forms.

In creating new forms of food and farm production, we must learn from our past and how our position in the world has been determined by what we eat. We need to promote the role of food and farming in more Brexit debates, and to do so we need a vision of the food and farming system we want in the future, one that is better not just for ourselves but also for the planet. Food and farming are a great metaphor for what we want in the rest of our society, as we often say 'we are what we eat'.

We have a glorious opportunity to set matters off in a new direction. Whatever happens, whatever sort of Brexit, or whether – even – we stay in, the issue of the present CAP subsidies will be up for grabs. In this book, the suggestions for how we could better use those subsidises, presumes we are Brexiting. However, if we stay in the EU, it may be feasible to make changes, as a process of CAP reform in the EU is underway.

Whatever we do, the state can play an important role in building better rural communities, and policies that deliver better for the land and labour. You have a great chance to join in the debate and use this book to identify the main battlegrounds. Argue with what I say, and make up your own mind – as to what is best for most of us, for the future of this country and for the earth. Then go out and fight for it

While I hope to have captured the essence of Brexit, food and farming, changes will inevitably occur, so I refer you to the website www.bittersweetbrexit.co.uk for updates and chat.

Photo: 2 The author is looking forward to taking ideas in this book on to union education courses.

PART I

The State We're in

1
All Change

This chapter sets out where we are now, and the significance of food and farming, and how we may be undergoing some of the biggest changes in our food habits for many years. There is an opportunity to discuss what we want from our food system in ways we have not been able to do for decades. But in order to decide where we may be going, we have to look at where we have come from to learn some lessons.

In June 2016, a small majority of British people voted to leave the European Union. This is going to mean one of the biggest changes in our food and farming system in the last 200 years. The Tories repealed the Corn Laws in the 1840s and Labour introduced the Agricultural Act following the Second World War. Now Brexit takes us in another direction.

When we voted, there was virtually no discussion about how leaving the EU would change what and how we eat, or indeed much else. Yet, the changes will affect virtually every part of the food and farm chain. Our exit will change how much we pay for food, who works the soil, what we eat, how our land is used, and most importantly, where we get our food from. It won't just be a matter of losing the directives, changing the laws, deciding what we do about tariffs, but more a matter of who we want to be.

In the run-up to the Referendum, there were some rumblings about farm subsidies. Farmers were reassured that something similar to the EU subsidies would continue. But days after the result, government ministers were saying they couldn't promise anything. So alarmed were farmers that the Chancellor, Philip Hammond, guaranteed a month later that farm subsidies would remain till 2020. The Tory manifesto for the June 2017 election extended that to 2022. That is no time at all. When I lived on a Lancashire hill farm in the 1970s, we had to plan way beyond that timescale.

The food and farming sector voted in different ways. Fishing communities voted overwhelmingly for out – over 90 per cent. The National Farmers Union (NFU – the big farmowners' association),[1] was

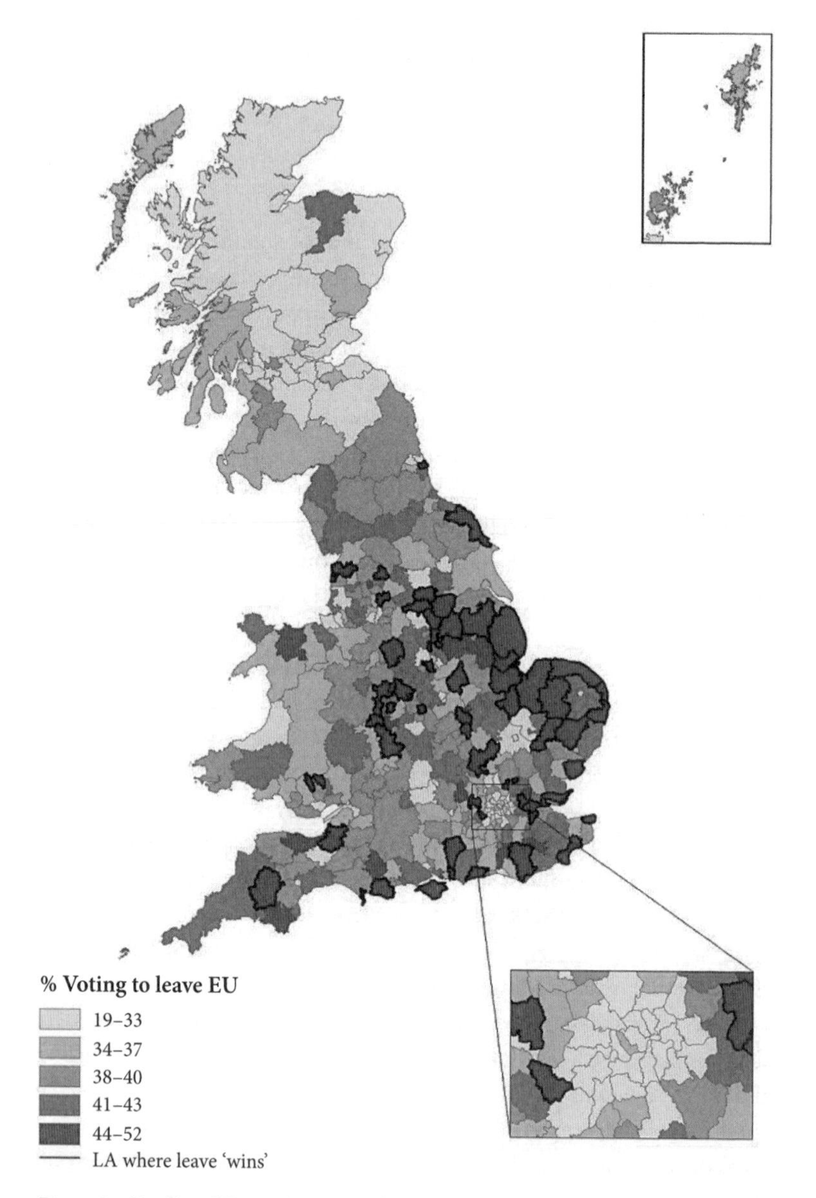

Figure 1 Predicted Brexit Vote, April 2016, predominantly around arable land in eastern England. (http://blogs.lse.ac.uk/politicsandpolicy/can-we-really-not-predict-who-will-vote-for-brexit-and-where/)

divided, some being dependent on migrant workers, others on subsidies. The Fresh Produce Consortium adopted a policy of neutrality. 70 per cent of members of the Food and Drink Federation wanted to remain. The Country Landowners and Business Association (CLA) lobbied ministers to 'do the right thing', but didn't take a position.[2] Farmers in Wales were convinced their subsidies would be safe, so voted to exit. I thought it was like turkeys voting for Christmas. Small farm tenants were clear they did not want to exit, as they believed the EU best served their interests. Tate & Lyle were avidly Brexit.

I spoke at a Food Ethics Council/Kindling Trust Conference in Manchester, one of the few food conferences addressing Brexit, in the run-up to the Referendum, where I made the case for remain. While I thought the EU's Common Agricultural Policy (CAP) was crap, I feared that our land and labour would be less protected outside of the EU. These two aspects are important to me as I believe that all our wealth is created from the soil and labour. Aspects of food production are of particular concern to me as I've represented farmworkers and I am a soil zoologist. I gained a doctorate in the early 1970s in soil ecology at Wye College, London University. I was on the National Sector Committee for Rural and Agricultural Workers of Unite, the union, having also represented them for many years on the Health and Safety Executive.

At that conference, I spelt out what drove Brexit in the first place. The main thrust to leave from the public came from the East of England, driven primarily by concerns over the number of migrant workers in the fields. I see these field operations as 'plantations' as they run monocultures with migrant workers on vast tracks of land – the definition of a plantation. While many of us think we are helping the country by buying vegetables wrapped in a Union Jack, we don't want to look too closely at how these crops are grown. Work conditions are so bad, most of us won't work there, and the soil conditions are also being badly degraded. We'll see more in the chapters on labour and land. People living in these growing areas feel that their own culture is being pushed aside, and their voice not heard.

During the talk, I put up a map of the predicted Brexit vote (see Figure 1), alongside another map showing the types of farming in Britain. The two maps were closely matched in terms of likely Brexit vote and areas of ploughed – arable – land in the East of England. In the Referendum, this Brexit vote was joined by high Brexit votes in Northern towns, where I have lived for 30 years. In these Eastern arable areas people voted 3–1 to Brexit, while the Northern mill towns voted around 2–1. Those in the cities, oblivious of the ways we produce our food, happy just to have cheap, convenient, fresh food, wanted to Remain.

That original thrust came because many people in the East of England didn't like the way we are producing our food. Good honest migrant workers doing jobs we won't do at all, in all weathers, were seen to be 'taking over our culture'. The blame was put on the migrant workers rather than the mode of agricultural production. The surrounding communities felt they had lost their own identity – just as many people in the Northern mill towns feel. We cannot duck this issue. It needs sorting, and is crucial to how we produce our food in the future.

While acknowledging that immigration was a major component of the Brexit vote, how different the outcome might have been if people had not also been under the cosh of austerity. Since the banking crash, the Coalition government of 2010–15 and the Tory government since have cut funding to schools, hospitals, bus routes, libraries and welfare benefits, mainly in areas outside the wealthier parts of London. The list goes on, with shortcuts in safety and health, frozen public-sector workers' wages, final salary pensions closed and wages in the private sector held back in an effort to plug the pension hole. We call this 'austerity', but it hardly begins to describe the pain.

Investment levels are at their lowest since the Second World War, and debt is higher than before the crash. During the June election of 2017, we kept hearing the Tories saying, as did the Prime Minister on 'Question Time',[3] that 'there isn't a magic money tree we can shake.' Yet I know of an orchard full of magic money trees. It is a walled garden in the City of London, where only bankers are allowed. There they grew these magic money trees, producing nice juicy fruit – called bonds. Reaping this fruit harvest is called 'Quantitative Easing' (QE). This is where the government pays for these fruits to the tune of £375bn, yes *billion*, as explained by the Bank of England[4] and an organisation called Positive Money.[5] No harvest is too abundant here in the orchard of magic money trees. They know that if there is ever another 'banking crisis', the government will come and pick, pluck and pay for their juicy fruit, all over again. That must be where the phrase 'rich pickings' comes from.

In financial jargon, the Bank of England created new money (electronically or out of thin air) to buy back bonds from the commercial banking sector (ordinary banks to you and me) which was then free to use the proceeds as it wished. According to the *Financial Times*, in an article called 'There is a money tree – it's called QE', QE boosts assets like property and equities. So the already rich do well with their stocks, bonds and buildings, while the poorer and younger have to pay higher rents and don't get a final pension settlement anymore. It means there are hidden tax cuts for the rich but hidden tax rises for the poor. The *FT* went on to ask: 'QE may have contributed to the rise of populism. Could Brexit, Trump, and the dissatisfaction across western nations, be partly due to its effects?'[6] So while we were bailing out the bankers, many were blaming the immigrants for their ills.

In this book I want to give you a flavour of a different world, where we could create a different food and farming system, which produces fruit for us all; lovely, lush, fresh, delicious and excitingly new foods to feed and nourish us all. It is quite feasible.

Up for Grabs

Perhaps Brexit will give us the chance to change. For a year after the Referendum, all we heard was 'Brexit means Brexit', whatever that means. We were told 'no deal is better than a bad deal', whatever that means. And that the government would get the best possible deal – whatever that means. It was as if everybody knew what they had voted for, and there was only one interpretation. There would be no discussion – the 'people had decided'. It took one individual, Gina Miller, to force the government even to allow debate in the mother of all parliaments. It all looked set. With a clear majority and a statutory period for the parliament, the 'will' of the people would prevail – whatever that meant. Yet, it seemed this invincible position was not enough – the Prime Minister, Theresa May – wanted a mandate for the negotiations, as a strong and stable leader. She called a snap general election.

She didn't get a mandate. She got the opposite. Overnight everything was up for grabs. The clear 'hard' Brexit we had been heading for, looked a lot less clear. All politicians were asked to clarify what they understood by Brexit, and where they stood. People started asking what is a 'hard' and a 'soft' Brexit? The negotiators in Europe said they didn't recognise either term. Some – like Michael Heseltine – raised the possibility that perhaps we could get what we want by staying in the EU. Others, like Ruth Davidson of the Scottish Conservatives, said they weren't bothered about coming out of the Single Market but did want to go off around the world doing deals. Labour fudged by keeping quiet on the matter, thereby attracting Remoaners in the South and Brexiteers in the North.

What is clear, amid all this confusion, is that we now have the opportunity to influence matters over the next few years in all sorts of ways. In this book, I will point out the options regarding food and farming. This requires explaining many of the issues involved with any form of Brexit, and I hope to cover most of those directly affecting food and farming.

Once outside the Single Market, we are out. Yet many believe that we can still have the same market access. It's called 'wanting our cake and eating it'. The response from Germany invoked another foodie phrase – that we cannot have an '*a la carte* Europe'. The process of leaving the Single Market – divorcing from our other 27 partners – will be complex. Extricating ourselves from all sorts of laws and institutions will be hard enough, but there are many more problems ahead if we leave the Customs Union.

Customs is about taxes and food trade. Sorting out all the food taxes is a mammoth undertaking. Some 2,000 agricultural products attract taxes on trade (tariffs), each complicated with quotas. There are a further 15,000 processed foods (PAPs) attracting tariffs. Angela Merkel said that there will be no 'cherry picking'. The EU says we have to be divorced before we can talk about trade deals.

This book gets its name from the definition of bittersweet: (of food or drink) sweet with a bitter aftertaste, as in: 'she sipped the bittersweet drink'. Some say there is a land of milk and honey ahead – the sweet version. Others say that we are about to throw ourselves over the Tariff Cliff into the waves below. We know there will be bitter twists – barriers, blockages and borders. There will also be opportunities and challenges ahead. This book sets up debates about what is important in food and farming in terms of policy, politics and parliament in a way we've not been able to do in the last 50 years.

Many food businesses will be worried. The dependence on migrant workers from the EU is now clearly in jeopardy. Food producers have come to rely on easy access to the Single Market, moving food ingredients around freely. Lorries can come and go now, but in future there will be checks at the borders, whether for health or tax reasons. Yet we keep hearing people saying they want free access.

Add to this the hidden protection that has been afforded to food-producers and farmers. It has been a deliberate EU policy to protect food and farming. That used to be 'our' protection, now it is 'theirs'. Many food imports are subject to taxes – tariffs. Once outside the Single Market and Customs Union, many will urge that we should have access to cheaper food from abroad. For example, the *Economist* says, by protecting food producers, UK customers are being 'milked'. For producers there will be a double whammy as once we step outside the EU, we will be considered a 'third country'. Not only will the protections go, we will be subject to tariffs when we want to export our food into the EU – our biggest market.

Despite all the talk about 'the best possible deal for all', there is a major contradiction. Reducing tariffs on foodstuffs coming in to the UK would make food cheaper for consumers. But that puts the food and farm producers in the UK under increased pressure – when they have been protected up till now. Clearly there is conflict. Each foodstuff has different rules, tariffs and quotas, and there is no sign of how these will be sorted out. We will pick up on this in Chapter 2. First, let's take a closer look at where we are now.

WHERE WE ARE NOW

At present, we barely produce half of our own food (54 per cent to be precise[7]). We import twice the value of food that we export. The Food Trade Deficit (FTD, as I'll call it, is $33bn). These values are expressed in dollars, as that is the main currency of international trade, and the import/export figures are recorded in dollars. It is useful, as import/export figures in sterling are drastically different from a year or so ago, as the pound has devalued since the Brexit vote.

By spending the money spent on imports on our own foodstuffs, we could recycle money back into our economy. The amount we would spend on food would be the same. If we spent half of what we import on our own produce, our producers and the local economies would be a whole lot better off. Instead of going all over the place, often into speculators' pockets, we could spend it on ourselves. We could do a lot with that sort of money – over $30bn.

We spend less than 10 per cent of our earnings on food. The rich spend slightly less, and the poor a somewhat higher percentage of their earnings. This is the smallest proportion of any of the EU countries. More than any other country we have got used to 'cheap food', relying on foodstuffs coming from other countries. That cheap food costs the earth in terms of resources, water, energy, pollution, global warming and land used, all of which we will explore later. Over 70 per cent of our environmental food footprint occurs abroad – i.e. we do twice as much damage to the environment in other countries as we do in our own, because of the way we produce our food.

It is also clearly damaging our health. We are the fattest country in Europe. In percentage terms, the UK has the highest rate of women's obesity, while men manage third place, behind Malta and Slovenia. Obesity is a global epidemic. While I was working for the World Health Organisation (WHO) developing their education materials on food and nutrition plans and programmes, we asked participants, from across Africa and Asia, how their eating habits had changed in the last 25 years. They all said that they had moved away from sitting down to have a meal, to eating on the move. We found obesity and hunger alongside each other, even within families.

Obesity accounts for about four out of five cases of type 2 diabetes – the one you develop rather than are born with. It is caused when insulin, secreted whenever we eat sugars or refined carbohydrates, stops working properly. It is estimated that type 2 diabetes is now responsible for a tenth of all NHS costs. The total cost (direct care and indirect costs) associated with diabetes in the UK currently stands at £23.7bn and is predicted to rise to £39.8bn by 2035.[8] (See Chapter 8.)

The proportion of what we produce ourselves to what we eat (self-sufficiency ratio) has been going down from a high of around three-quarters in the 1980s to just over half now. We can mark the decline from the moment that Mrs Thatcher opted to leave food policy to the supermarkets. Now could be the time to turn that round, and say we could easily produce more food for ourselves, and in the process build thriving new businesses.

Imports/Exports

We export about $33bn in foodstuffs, animals and vegetables. Around $8bn of those exports are whisky and gin. Included in overall food exports are $8bn of animal products, while fish in various forms account for a fifth. Vegetables bring in a further $3.3bn in exports, of which 15 per cent is wheat and grain – the same as tea and coffee. We manage to export $0.5bn of tea and coffee, despite growing neither.[9]

The government wants to increase food exports by £3bn (UK figure in £) over a five-year period. This 'ambitious' international plan was announced in Paris in October 2016. It sets out to increase drink exports – particularly beer and whisky to the US, Mexico and Australia – along with tea and biscuits to Japan, where apparently they have a taste for our afternoon teas.

We import about $66bn worth of *all* food – foodstuffs, animals and vegetables – around half of that from the EU. So we import twice as much in the way of foodstuffs, animals and vegetables as we export. The UK is the fifth largest importer in the world, and food accounts for about a tenth of the total.

In terms of foodstuffs alone, we import $34bn: wine accounts for 14 per cent of the total! Add in chocolate and those two pleasures account for over $6bn in imports. We can now grow grapes as far north as Morecambe. If only we could grow cocoa! Further we import $15bn animal products – again around twice as much as we export. Of the $15bn, $2.5bn is cheese, poultry $1.5bn and around $1bn each of beef, pig meat, preserved meat and fish. Despite producing more varieties than the French, we import two-thirds of our cheese from France. As Liz Truss, then Secretary of State for DEFRA, famously said: 'That is a disgrace.'[10]

We import $17bn worth of vegetables. That is five times the amount we export! And we are a vegetable-growing country. Grapes account for a billion of that. Citrus and bananas account for another $1.5bn. We import over $0.5bn worth each of tomatoes, apples, pears and other fruit, wheat, live plants, corn and frozen vegetables – all of which could be grown here. That could save £4bn. Imagine what we could do with that money.

Instead of exporting to others we should supply food to ourselves. We have a £200bn food industry employing one in eight of the UK workforce, all of whom could use those foodstuffs. Surely it makes more sense to spend time and money *reducing* food imports by £1bn to match the expected increase of £1bn in exports, as a way of reducing the Food Trade Deficit. Controlling our own borders' could mean controlling the food coming in. By selling food to ourselves, we are not only creating a vibrant economy, we are building some 'resilience' and sustainability into our food provision.

We rely on European countries for over half of our food imports, 27 per cent of our total food consumption. We also send most of our food exports to the EU – especially meat. As we move away from the EU, it is hard to see us maintaining that. Somebody on 'Farming Today' described it as challenging. I would say it is 'pie in the sky'. There are likely to be all sorts of taxes on food trade, and even if there are not, there will still be plenty of new border controls.

Markets

There is a massive problem with obesity and a massive problem of food waste, yet people still keep saying the problem is overpopulation. Simultaneous obesity and food waste show the problem is overproduction – not overpopulation. The present EU subsidies confirm this, as they pay out 40 per cent of the total EU budget to landowners not to produce any more food. For the last 40 years, EU farming policy has been trying to get to grips with overproduction of food. Not overpopulation.

This is because the Western world still believes in a market-based system. Markets are there to reward products in short supply. The converse is that markets don't reward abundance – when there is a good harvest. I am not the first to notice this – the bard of Stratford did, over 400 years ago. The Porter in Shakespeare's *Macbeth* imagines himself opening the gates of hell to 'a farmer that hang'd himself on th'expectation of plenty' (Act II, Scene 3).

Free markets act against increasing food production. The law of the market is the more there is, the lower the price. In the financial press you can read about bad harvests pushing up prices, as happened in the spring of 2017, when there was a shortage of Spanish lettuces and courgettes. So, it is obvious that good harvests bring prices down. The EU has tried all sorts of ways to mitigate this, yet still clings to the myth of a free market. When prices fall, production costs must be trimmed. This leaves many in the food chain badly paid and often hungry.

We must be able to develop a new farm and food system, without bowing down to the god of free markets. We may now have an opportunity to provide a healthy, varied and more sustainable food supply and develop our own farm and food system to the benefit of rural and urban communities alike.

This will not be just a matter of a few laws or taxes, but of who we want to be, where we want to go, and what we want to achieve. We will need to work out better our own identity through food and farming, and we will pick up on this in the last chapter. Rather than leaving matters to be decided in the finance houses, let us do so in the green fields of Britain. However, before setting off, it is worth looking back at how our food and farming has evolved over the last 200 years, to see what lessons we can learn.

HOW WE GOT HERE

Thomas Malthus in *An Essay on the Principle of Population* (1798) said manufacturing would never increase the wealth of the nation because food production was its primary economic purpose. An industrial worker

> will have added nothing to the gross produce of the land: he has consumed a portion of this gross product, and has left a bit of lace in return; and though he may sell this bit of lace in return for three times the quantity of provisions he has consumed while making it ... he cannot be considered as having added by his labour to any essential part of the riches of the state.[11]

How wrong he was. Within 40 years, Karl Marx was able to say that manufacturing had rescued many workers from rural isolation, as the Industrial Revolution swept all before it. Malthus was also wrong when it came to his more famous prophecy that population always outstrips food resources. The opposite is the case now, as we produce much more food than we can eat.

Corn Laws

In the early 1800s, following the Napoleonic Wars, the Corn Laws were introduced to 'protect' UK grain producers from cheap grain imports from colonies like Canada. The Corn Laws were hugely controversial. Free-market industrialists wanted the laws removed so they could access cheaper grain and thus produce cheaper bread – and pay their workers less. The Conservative, William Pitt, repealed the Corn Laws in 1846. Although the effects were not immediate, within 30 years British agriculture was in dire straits. It was unable to compete with cheap grain shipped in by steam ships, and meat coming in freezer ships, built in Northern Ireland. This marked the ascendancy of industry over agriculture.

The 1840s also saw the Irish potato famine, when an estimated million people – one in eight of the Irish population – died, while twice as many emigrated to America. At school, I was so concerned I avidly read about the 'culprit' – potato blight, a fungal disease called *Phytophthera infestans* – and was determined to help stop this sort of thing ever happening again. It wasn't till much later I found out that during the so-called famine there had been no shortage of grain to feed the gentry. There was sufficient food in the country, as corn was imported from India, from the beginning of 1848. The government stopped a soup-kitchen scheme that fed 3 million people daily.[12]

I wonder what Malthus might say today about the finance sector and whether it contributes to the 'riches of the state'. Obviously the finance sector contributes to the riches of a few. In my union, I remember sitting in the annual conference, with the rural and agricultural representatives – about 20 people but outnumbered by the finance contingent of around 200, all good people. I couldn't help but wonder whether their work added to the diversity of life in the soil, the biodiversity of life above it, or the living conditions of people working that soil. I suspect that when we discuss what sort of food system we want, the finance sector will want 'cheap food' to feed society – just like the industrialists of 150 years ago.

By 1875, British agriculture was in the doldrums and stayed that way for the next 75 years. In the Pennine hills, where I live, you can see many derelict farms dating from that period, as the inhabitants abandoned the land to find work in the mills. At the start of the First World War, we produced only a quarter of our own food. Promises were made that the land would be fit for heroes when the soldiers returned, but nothing was done to make rural life any better.

In the 1930s, quotas were introduced to prevent overproduction for various foodstuffs – particularly wheat, milk and potatoes. These maintained decent prices for the growers. The Wheat Act of 1932 put levies on incoming wheat and quotas on milling wheat grown here. Similar systems were introduced for milk, setting up the Milk Marketing Board to guarantee prices. Quotas were also introduced for hops. The government invested in land-based research as it considered it too important to entrust it to anybody else. The Marketing Boards were closed when we went into the European Economic Community (EEC, now EU) as they did not accord with the 'laws of the free market'.

After the Second World War

After the Second World War, both the UK and the rest of Europe invested heavily in food production. This was as a result of the threats to supplies during the war. We had depended heavily on grain from Canada that had been vulnerable to attack by German U-boats. The Labour government after the war invested massively in drainage, fencing and fertilisers and set up the National Agricultural Advisory Service (since privatised) to link research to farmers. In many ways, this was as significant as setting up the NHS, although it doesn't get the same acclaim.

A system of guaranteed minimum prices ensured that UK farmers received their just reward and helped to expand production. This central plank of postwar reconstruction made us less reliant on others for our food. It was one of the few periods in the last 200 years where the government intervened – a Labour government. To build home-grown crop production, guaranteed prices were paid for staple agricultural products. Clement Attlee's Labour government passed the 1947 Agriculture Act. It was introduced by an ex-miner, the then Minister of Agriculture, Tom Williams, with the express purpose of improving the UK's balance of payments by paying less for food imported from dollar countries. All parties supported the Bill. Its rationale was: 'The twin pillars upon which government's agricultural policy rests are stability and efficiency. The method of providing stability is through guaranteed prices and assured markets.'[13] The government instigated annual price reviews and prices were fixed for the main crops (wheat, barley, oats, rye, potatoes and sugar beet) for 18 months ahead. Minimum prices for milk and eggs were fixed for two to four years ahead. Farmers got the money, and speculators got nothing.

Throughout Europe, the Marshall Plan was in force. Named after the US Secretary of State George Marshall, the plan was partly motivated by humanitarian aims to end the misery of food shortages. The USA provided over £100bn (at current value) for Western European countries (26 per cent to UK), who spent about a quarter of that buying food, animal-feed and fertiliser – from the USA.[14] The plan helped the US to dispose of its grain surpluses and gave them the idea for foreign aid in the future. In the mid-1950s they introduced the Public Law 480 Act to provide cheap food, with long-term credit arrangements, for 'friendly' countries. This contrasted with foreign aid given by other countries, as their 'Food for Peace' programmes were 'in kind' – rather than real cash. 'The United States continues to rely on domestic purchases of U.S. commodities as the basis for its food aid programs.'[15] This was amended slightly in their 2014 Agricultural Act, as other countries said the USA was distorting markets. Europe did not want to remain

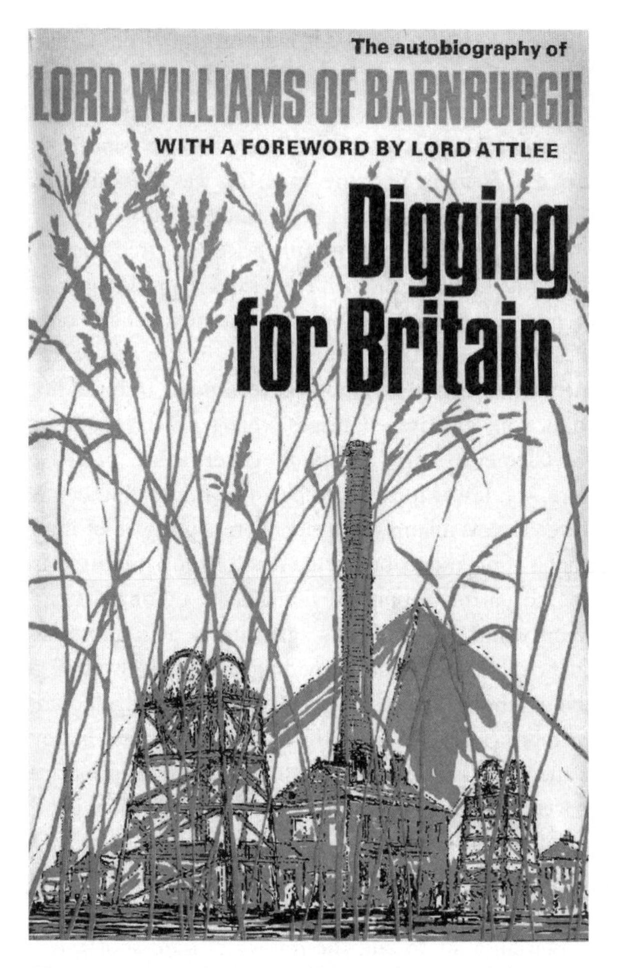

Photo 3 Digging for Britain was written in 1965 by the man who piloted through Parliament the 1947 Agricultural Act. Attlee said in the foreword that the Act 'effected nothing less than a revolution in British agriculture'.

dependent on US imports, so encouraged more food production in the 1950s, which helps to explain why the EU are still happy to pay out 40 per cent of their budget in agricultural subsidies.

When we entered the EEC, in the 1970s, our farm-support system changed and we junked the guaranteed price system, as the EEC believed in 'free markets'. However, the problem of overproduction, threatening farm prices, soon reared its head. The first attempts to limit farm production came in the late1960s, when Sicco Mansholt, then the socialist Vice President of the European Commission, proposed the first major reform of the Common Agricultural Policy to deal with overproduction. He suggested that 5 million small farmers give up farming. The plan was to redistribute the land, such that farms were big enough to guarantee owners better earnings. That plan went down like a sack of potatoes, so nothing happened.

My View

I was born just after the war and well remember rations, especially when sugar rationing ended. From the age of around twelve I was a nerd who wanted to 'feed the world'. Friends of the family confirm that I'd go on about growing algae in the sea to help feed the starving millions. We were bought up on 'waste not, want not – think of those in Africa'. I came up with the idea of collecting locusts to feed the world: after they had been killed, I planned to cover them with local cocoa and sell them on as a delicacy in New York. Nowadays that would be a model of sustainability, as eating insects (*entomophagy*) is now being proposed as a way to save the world.

I steered my way through one of the King Edward Schools in Birmingham, making sure I did biological sciences for A level in order to study agricultural zoology. I became one of the infamous 'Agrics' at Newcastle University, studying agricultural zoology – what we would now call 'Integrated Pest Management (IPM)'. We learnt ways to control pests and diseases, with chemicals only as a last resort. That degree doesn't exist anymore, despite IPM being important throughout the world. I wanted to do my bit to reduce pests and disease and thereby improve food production. I followed this with a masters in applied plant science at Wye College, before doing my doctorate. By the end I realised we do not need to keep producing more food. As the 1970s wore on, it became clear that we could produce more than enough food in Europe.

By the 1980s, overproduction of food in Europe was downright embarrassing. The EU had become a massive overproducer with butter and beef mountains and wine lakes hitting the headlines. We watched 'Live Aid', as the notorious butter mountain was on its way to reaching 1.4 million tonnes. The EEC gave some to Russia, slashed prices for pensioners and 'dumped' massive amounts on world markets.

While the 'Live Aid' planes took off, loaded with food to feed the emaciated bodies we saw in Africa, Eric Bogle sang:

Have you seen the children, who disturb our paradise?
Staring from the TV, with their empty dying eyes.
No trace of the anger, at the betrayal of the trust.
That left them to die like starving dogs, in the famine's bitter dust.[16]

The dreadful contradiction – between starvation and overproduction – became more and more obvious. But still people believe all we need to do is produce more food. We don't. We need to give the poor more money. The hungry millions don't feature in food markets because they can't afford to buy food. If they could, the market would expand to feed them. The very people who help make the food, often cannot afford to buy it.

There are two dominant farming forms – family farming and capitalist plantations. Most food crops, grown for export, are now grown on the other side of the world from where they were first cultivated. This surprised me at first, as I thought – like many other people – that it was 'natural' that plants grew where they did. Yet it is not particularly 'natural': a complex mix of factors is involved. Bananas now grown in the Caribbean were first cultivated in Southern China and Indonesia, wheat came from what is now Iran, Iraq and Turkey, but is grown mostly for export in the US, and cocoa came from the Amazon but is now grown in West Africa.[17]

Britain has been a major player in this distribution, and Kew Gardens a key resource. Most of these crop movements are due to changes in the mode of production, mainly from family farming to capital-intensive plantations. My professor of tropical crops argued that the movement was due to escaping indigenous pests, yet I thought it was due to the movement of money and labour. While at Wye College, I was President of the Middle Common Room, representing some 200 postgraduate students, mainly from the Commonwealth countries, who told me what went on in the wider world. What a shame we don't have that resource today.

When I realised that the problem was how to provide better food not more, it was like being a Christian in church and suddenly realising I didn't believe in God. My whole academic career and trajectory was to help to produce more food. Luckily a Brazilian Archbishop, Dom Helder Camara, helped me out when he said: 'When I give food to the poor, they call me a saint. When I ask why the poor have no food, they call me a communist.' I like asking questions.

When I looked back at agricultural history, I realised I wasn't the first agricultural scientist to recognise this dilemma. One of the best British agricultural scientists was Sir George Stapledon, director of the Plant Breeding Station at Aberystwyth for 20 interwar years, who promoted the importance of grasslands and encouraged an ecological approach. To him, 'free commerce' was the source of the problems confronting agriculture. Britain was incapable of feeding itself, making it dangerously exposed in times of crisis. Stapledon argued that this was due to its having sacrificed agriculture for the sake of industrial growth and imperial security through 'free trade'.[18] Stapledon wrote: 'The possibilities of generously feeding the whole of mankind ... are adversely affected only by man's wilfully restricted capacity for breaking down the artificial barriers he himself set up by exaggerated deference to the free play of "economic laws".'[19]

After the Second World War, John Boyd Orr, the first Director of the Food and Agriculture Organisation (FAO), clearly understood that overproduction was the problem. Journalist Ritchie Calder asked Boyd Orr why 'he had a chip on his shoulder'.[20] He politely explained:

> half the population of the world suffer from lack of sufficient food while farmers suffer ruin if they produce 'too much food'. Adjust our economic and political systems to let these two evils cancel each other out … A thousand million peasants and farmers in poverty because they cannot produce the food the hungry need, or if they could, would face ruin because of something called 'overproduction'. The world, through science and common sense, could produce the food. Think of the dividend, not only in farming prosperity, but in human well being![21]

I came across Professor Spedding of Reading University. He said: 'Consider the facts that world agriculture could easily produce vastly more food, that over-production has been a major problem of many regions, that considerable efforts and incentives have been devoted to reducing production in many areas, and that a great many people go hungry … it has to be recognised that, just like any other industry, agriculture is practised for a wide variety of reasons, to make money being one of the most common.'[22]

Many agricultural scientists turn a blind eye to the effects of free trade, so I set up a group in the mid-1970s to look at the role of agricultural science in a capitalist world. This was while I was working for an organisation called the British Society for Social Responsibility in Science (BSSRS). The President of the BSSRS was Maurice Wilkins, one of the three scientists awarded the Nobel prize for mapping DNA. We called ourselves the 'Agricapital Group',[23] which has been referred to as the first 'politics of food' group. We produced a leaflet called *Our Daily Bread*.[24] Associated British Foods served us with a writ, saying that they 'didn't altogether agree with our analysis of the British bread industry, in particular the description of work hazards in the bakeries'. We did not republish.

While working as 'pollution man' at BSSRS, I helped uncover one of the most important new industrial health issues in the modern age – that the manufacture of PVC could cause a rare cancer among the workers involved. So, with help from 'Science for People' groups throughout the country, we developed a magazine called *Hazards Bulletin*. It is distributed throughout the trade union world and I wonder how many lives and limbs the *Hazards* campaign has saved. It will be even more important as we come out of the EU and loosen controls on substances hazardous to our health in workplaces or pesticides on farms, which we will pick up in Chapter 9.

Photo 4 The author is delighted that the *Hazards Bulletin* – now magazine – has played an important part in the movement that has saved so many lives and limbs.

While I was at BSSRS, I got to know a *New Scientist* journalist called Joe Hanlon, who later wrote a book called *Just Give Money to the Poor*.[25] He said if you give people money, you create a market from which they can develop. At present the hungry are the poor, and cannot afford to access the market. They starve not because there isn't the food, but because they can't afford to buy it. If they could, the market would grow. So give them the money.

Overproduction

While we may be concerned with the starving millions, the owners of food capital are obsessed with the opposite – what to do with too much food. This is the continuing problem for food markets – yet we don't talk about it. We cannot feed the voracious appetite of capital expansion. We cannot eat enough to feed its hunger for accumulation. Capital is obsessed with saturated markets, not saturated fats. Humanity and capital have completely opposite concerns. It certainly shocked me.

Throughout *The Ragged Trousered Philanthropists*, published in 1914, Robert Tressel quotes from *Merrie England*, whose author, Robert Blatchford, asks 'Can England Feed Herself?' He answers: 'This earth this day produces sufficient for our existence, this our earth produces not only sufficiency, but superabundance, and pours a cornucopia of good things down upon us.'[26] He goes on to say that, if we in England used all available land, we could feed 200 million people.

We have more than enough food to feed 14 billion people now – twice the global population. Remember one-third of all foodstuffs go to waste and over a billion people are obese. We can easily produce more; the problem is selling it. The iron law of any market says that when any commodity is in plentiful supply, the price drops. Scarcity brings higher prices. Because prices go down when producers do well, growers run on a treadmill of producing still more, more efficiently, to make up for the loss of profitability. This dance of efficiency is at the expense of sustainability, biodiversity and global warming.

Look at the dairy industry in the UK, where milk prices no longer cover the costs of production. As dairy farmers become more 'efficient' it exacerbates the situation. It is a treadmill with nowhere to go. Some nine UK retailers worked together to protect their dairy farmers, by agreeing to pay them a 'living' price. They were thanked for their efforts by being taken to court by the god of free markets – the Office of Fair Trading. Four major retailers, including Tesco, were fined £50m for 'collusion'.[27]

Overproduction is the norm in world trade too. The dominant food overproducer in the world is the USA. US farmers were urged in the 1970s to produce 'row to row', meaning they should grow as much as they can, by their Secretary of State, Earl Butz. To get rid of the excess corn and grain, he arranged for it to be sold to Russia at the height of the Cold War (the Great Grain Robbery), increased cornflakes exports, went to Japan to learn how to make corn syrup (fructose), and encouraged intensive cattle-rearing.

Then in the late 1980s, a certain 'Milk Snatcher' decided food policy was better 'left to the (super) markets'. Margaret Thatcher started selling off vital agricultural support assets – like the Plant Breeding Institute (PBI) in Cambridge to Unilever. Monsanto later bought PBI from them. Bayer is buying Monsanto for $66bn to form one of the three largest agribusinesses in the world.[28] Syngenta (which owns all ICI's old patents) has just been bought by China, which we will investigate later.

By the early 1990s, while hunger still stalked the world, the EU had to get to grips with its beef mountains and wine lakes. By then, the farming community had less political influence, while the green non-governmental-organisations (NGOs) had more. Recognising that they didn't need to produce as much food, the then EU Agricultural Commissioner Ray MacSharry introduced some reforms (1992) to the Common Agricultural Policy (CAP). One such reform was 'Setaside', where farmers were paid *not* to produce any foodstuffs.

The MacSharry reforms also offered generous Arable Area Aid (AAA) payments. Farmers got around £150 an acre if they went arable – i.e. ploughed it up. Within a few years, English counties like Essex, for centuries a beautiful patchwork of mixed farms with pasture and ploughed land, were emptied of their livestock and became the forerunners of today's plantations. We will see in Chapter 6 what a disastrous impact that has had on our soils and how important it would be to bring back mixed farms.

Many of us used to sit down and have a meal with colleagues at lunchtime, usually for an hour. Now we graze at our desks. We lost that hour sometime in the 1990s. We knew that stopping our tea break was the one thing our union members would unite and fight over. Yet our lunch/dinner hours have just disappeared. The subsidised canteen often provided our main meal of the day. Local women, who cooked and served local food, were replaced by 'cook chill' machines dispensing dross, and facilities to heat processed food. When running World Health Organisation (WHO) food and nutrition courses in Africa, we found the same had happened everywhere. I helped run a course for IKEA about 'responsible' food, as they are one of the top 20 restaurateurs in the world in terms of the number of meals they serve. It was noticeable at their Swedish headquarters that co-workers sat down to lunch together, recognising it as an important part of their working day.

School meals also came under threat. The early Labour Party pushed the Liberal Party in 1906 to introduce the first school meal acts, when it was realised that workers were malnourished and not fit to fight in the Boer War.[29] However, in her first year as Prime Minister, Mrs Thatcher abolished the minimum nutritional standards for school meals and removed local authorities' statutory duty to offer meals. Children dashed off to the chip vans. Jamie Oliver in 2005 tried to turn that clock back. In 2017, 98 per cent of primary schools offered hot meals as part of the Universal Infant Free School Meals schemes pushed by the Liberal Democrats in Coalition. The Tory manifesto proposed to abolish school meals, replacing them with free breakfasts, which were said to be 'more cost effective'. I dread to think what those breakfasts might have consisted of. Around 900,000 children from families who are 'just about managing' would have been affected. Jamie Oliver said he was 'shocked and disappointed'. The Tory government dropped the proposal in the Queen's speech, June 2017.

In the 1990s, more women joined the workforce than ever before. While we cheered for equality, it was, in part, to make up for the decline in the amount the male 'breadwinner' bought home. Unions had lost bargaining power in the 1980s, so women made up the difference. As a result, the manufacture of processed and packaged meals increased dramatically, providing a ready-made replacement to home-cooked food. So-called 'women's work' – cooking – was replaced by factory-made meals.

The 1990s were also the golden days of globalisation. Up until the 1990s 'futures contracts' had been used to help farmers deal with the uncertainty of growing crops, such as unforeseen weather conditions. A futures contract allowed farmers to sell their crops at a future date at a guaranteed price. However, following pressure from US financial regulators, these contracts can now be bought and sold by speculators who have no interest in the actual food being traded. By buying and selling the contracts, they could profit from prices changing over time – i.e. betting on the price of food. This 'neoliberalisation' may have helped global food trade grow and around the world, standards of living were rising. But there was trouble brewing.

The World Trade Organisation (WTO) was set up in 1995. This members-only club promoted multilateral trade liberalisation. However by the early 2000s, their next round of global negotiations became a saga of missed deadlines. The major developing countries, led and represented mainly by India, Brazil, China and South Africa, thought that the subsidies and tariffs that the EU and USA in particular applied to control food production contradicted what the WTO is supposed to stand for – free trade. The US and EU each subsidise farms at a cost of around $50bn a year.

In the late 1990s, the EU still had to deal with its food/farm surplus. So a new system of subsidies was introduced in the 2000s – called the Single Farm Payment (now Basic Payment Scheme). This pays farmers for 'looking after' the land. It means that if you own more than 10 acres, you receive £100 per acre per year from the EU. The more land you have, the more you get – for doing nothing. Obviously, this mostly benefits big landowners. Like the Queen. She gets a third of a million pounds a year for her land in the Trough of Bowland in Lancashire. Her neighbour, the

Photo 5 Under the Common Agricultural Policy large landowners scoop up £3bn annually. They include the Queen who pockets a third of a million for the land she owns in the Forest of Bowland. Credit: Mark Harvey

Duke of Westminster (or rather the Grosvenor estate), gets even more per year from CAP. Tenant farmers complain that the subsidies don't get passed on to them. This subsidy amounts to £3bn/yr. What we do with these subsidies in the future will speak volumes about what we want to do with our food. We will look into that later.

By 2008, the EU was once more facing its perennial problem of overproduction. The European Commission pledged to buy up 30,000 tonnes of butter from farmers across the 28-country union and over 100,000 tonnes of skimmed milk powder, for £230m, storing it in silos across the continent, which already contained over 300,000 tonnes of unwanted sugar, 16,000 tons of unwanted maize and wheat and millions of gallons of wine.[30] Nigel Farage, then UKIP leader, said that 'the return of the butter mountain was proof of the lunacy of the Common Agricultural Policy – another shocking waste of money.' He forgot to criticise the US for paying out similar amounts of subsidies.

Then the financial crash of 2007–8 shook the world. The banks had been making money out of selling debt, until somebody realised the King was in the altogether. We are still paying for the government's bailout of the bankers. The massive financial crash was associated with a food crisis. Some tried to explain all the food price hikes on floods and droughts, as if it was all due to natural causes. But economists noted that food had become a better investment than mortgages, turning the financial crisis into a food crisis.

Higher food prices hit countries in the Middle East – despite the food mountains in the EU. This led – in part – to the Arab Risings.

> Free Trade Agreements and the pressure to join the WTO made the Arab world an easy open market for Western-based multinational food corporations ... Field studies in Lebanon, Yemen and Jordan indicate that for those Arab countries where there exists a potential for agriculture, dumping of subsidized food has contributed to the demise of the local farming sector ... No wonder farmers chose to opt out from agriculture and to migrate to the cities where they become net food buyers.[31]

These Middle East countries had become more dependent on imported food – to around the same extent as we are now. So they were subject to the vagaries of global food prices – and global gamblers. The vegetable seller, Mohamed Bouazizi, who set fire to himself late in 2010, lit the fuse for the Arab Risings.[32]

Since 2007–8, we have opened nearly 2,000 food banks in the UK,[33] some 1,300 run by the Trussell Trust and 650 by independent organisations.[34] These food banks are now part of our society. Inequalities have grown to the point where many of the people who use these food banks are in work – but so poorly paid that they cannot afford to buy food in conventional retail outlets. Judging from our local food bank in Burnley, it is the big local manufacturers who provide foods such as biscuits, bread and brown sauce, which would not pass school nutritional standards.

During a Commons debate about food banks, the then Welfare Minister, Iain Duncan Smith, who lives on a £2m country estate, left early, leaving his junior minister, Esther McVey, trying to blame Labour. Yet the number of people going to these banks has risen ten-fold since 2010, from 4,100 to 400,000+.

The biggest contradiction of the British food system is that while desperate people have no alternative but to go to food banks, food waste is obscene, and the rate of obesity continues to rise. It has risen steadily for the last 30 years, coinciding with the 'leave it to the supermarket' philosophy. There, the customer wants 'choice' – but the choice is between cheap, crappy foodstuff and more cheap, crappy foodstuff.

The contradiction that I realised all those years ago – that the problem is not overpopulation, but overproduction – has still not been addressed. We need to produce 'better, healthier and greener food'. And we can. Leaving Europe may be our opportunity to do so.

But it will be a battle. Consumers will still want cheap food. That won't stop any time soon. Yet cheap food costs the earth. I believe there are ways to square this and these will be explored in this book. We cannot rely on individual consumers to do this. If ever there was a case for state intervention, this is it. It is not a matter of being a 'nanny state' – although the better off in this country are more than happy to have nannies look after their children. It is the role of a progressive, forward-looking state to do things no individual can do. It means we have to have political answers, not individual ones, however well-meaning.

In spring 2017, the Prime Minister, Mrs May, said she wanted the UK to be a world leader. Perhaps we can be, but I do not want to see us as a world leader in flogging more booze round the world. We should be demonstrating to the world that our food service is healthier, greener and better for those who work in it.

We need to put the role of food and farming centre stage. My fear is that the Gherkin (the iconic tower representing the finance sector) in the City of London will get all the attention and all the money. I say that, as an Honorary Research Fellow at the Food Policy Unit, at City University, London – just around the corner from the Gherkin. We should be encouraging crops of cucumbers, courgettes , celery and a host of new crops. We also need to grow our minds to create new forms of food business and develop our sciences to look at a wider range of issues. We forget what is under our noses – food, because it is always there. But we should stop taking that for granted.

We have an opportunity for the £10bn farm sector and the £200bn food service sector in the UK to grow. Our fate is 'up for grabs'. This book covers all the options – the various sorts of Brexit, as well as the remote possibility that we might remain – exploring how to achieve the best for our food, farming, land and labour. It may be that this can be best achieved as the UK, but it may form part of a plan for a revitalized Europe. We could be *in* Europe but *against* some of the grosser idiocies. Throughout the book, I'll be asking what is better for our food and farming, being in Europe or out, and if so in what way. So let's set off and see what might happen to our food and farming if we exit Europe, and then what we do afterwards in terms of moving on.

2
Coming Out

IF, HOW AND WHEN ...

This chapter sets out some of the Brexit options, and how they relate to food and farming. In particular we look at the consequences of coming out of the Single (internal) Market and then the different decision to leave the Customs Union. The Single Market is all about harmonised rules allowing free access without any non-tariff (safety) barriers. The Customs Union sets up tariff (tax) barriers to anybody outside, which pose a whole different set of issues for food and farming.

TABLES TURNED

During the massive changes ahead, there will be (sweet) opportunities, and (bitter) challenges. The bittersweet rationale is to look at both the 'opportunities' and the 'challenges'. New options will open up, but there

Life can be bitter-sweet but at least you can get to control the sugar

are all those laws, treaties, directives and tariffs that affect every aspect of the food chain that went largely undiscussed in the Brexit debate. We – consumers, suppliers, providers, and all those citizens who need to eat – should now have those debates as to what opportunities lie ahead.

The EU is both a Single Market and a Customs Union. We have heard a lot about the first but little about the second. Theresa May said in her Lancashire House speech in January 2017 that her idea of Brexit was to leave the Single Market. She wasn't sure whether that also meant leaving the Customs Union. The Labour Party in the run-up to the June 2017 election said that the Tories should leave the exit from the Customs Union 'on the table'. I suspect very few people understood what that meant, when they voted. Let's look at that table.

Imagine the table top is the Single Market. We've put a table cloth on it to make sure the surface is even. All the barriers to trade have been swept aside so we can move the (harmonised) food, drink, water, knives, forks, glasses and plates around easily. The food is served to guaranteed safety standards. Our waitress/waiter may come from anywhere in the EU. We can serve ourselves whatever we like. The air is pretty clean, as we agreed not to pollute it. This is what goes on inside the Single Market. That is what most people understand as Brexit. After Brexit, we can make new rules for the cutlery, food standards, and who makes, cooks and serves it. That is what most 'Brexiteers' expect the Brexit minister to negotiate.

Around the edge of the table is the Customs Union. This Customs Union puts up barriers to foodstuffs coming into the Single Market. It does this in two ways. Customs offices collect a tax (tariff) on farm and food goods. They also impose standards to make sure the foodstuffs are healthy. Countries are allowed to set health barriers, provided they apply the same standards to their own produce as well as those foodstuffs coming from elsewhere. It could be a relatively low table, with low tariffs, like a coffee table. Roasted coffee has a small tariff (7.5 per cent). However, the EU has gone out of its way to protect its farms and food producers, putting up high tariffs (30 per cent) for foodstuffs like meat. Our meal in June 2017 is served on a high table.

If Britain wants to go off doing deals with other countries, we have to leave the table. In the Customs Union we are surrounded by 2,000 agricultural products attracting tariffs, and a further 15,000 for processed foods. Those who want to leave the Customs Union, so we can do deals elsewhere, I call the 'free traders' – to distinguish them from Brexiteers. Jiggling with the food and cutlery is one thing, jumping off the table is another. When we jump over the 'Tariff Cliff', we will find ourselves looking back at a high table for most of our meals.

It is possible to be outside the Single Market, getting rid of the laws, and stay inside the Customs Union, protected by those tariffs. Other countries have. There is another organisation all EU members belong to called the European Economic Area (EEA). The signatories are the 28 members of the EU and three of the four members of the European Free Trade Association (Norway, Iceland and Liechtenstein). Being a signatory to the EEA Agreement entitles an EEA member state to unrestricted, tariff-free access to the EU's Internal Market (the correct name for the Single Market!) for all products – *except* fisheries and agriculture!

The government talks of negotiating the best possible deal and sees carrying out the divorce process (Brexit) at the same time as doing a deal on trade with the EU. This is yet again an example of the 'cake and eat it' mind-set. The EU says we have to be divorced first before doing any deal. After sorting all this out, we can look to see where we are heading. There are two main currents – those taking us globally and those looking towards buying British – which we will pick up in the next chapter. Here we look at Brexit and doing deals, in terms of food and farming, to help identify the opportunities and challenges ahead.

BREXIT

Like any divorce, this is a matter of dividing up the assets and sorting out the previous commitments. The European Commission has said: 'Britain must pay to leave the European Union in the same way as friends going to the pub must pay for their round of drinks.' Margaritis Schinas, a spokesperson for Commission President Jean-Claude Juncker, said: 'It is like going to the pub with 27 friends. You order a round of beer but then you cannot leave while the party continues, you still have to pay for the round you ordered.'[1]

Our law lords have said that we wouldn't be breaking the law if we didn't pay up. But that would jeopardise any trade deals. The main thing we know about Brexit is that there are a vast number of unknowns. Disentangling the laws will keep lawyers rich for a long time.

We have spent years creating a 'level playing field' between 28 countries. The aim has been to work together and to create a seamless system that we all understand. This involved creating directives, which member states could adapt to local conditions, and regulations, which had to be followed to the letter. The first legal job is to decide which bits of the regulations can be transposed directly, and which bits cannot.

Key to the operation of the EU Single Market have been four freedoms – freedom of movement of goods (e.g. food), capital, services and people. Achieving this 'harmonised' system required all sorts of Institutions to be set up, to which the UK contributed. We made commitments to these institutions: reparations would be expected in any sort of divorce. We will have to replace most of these institutions with similar institutions. On top of this are policies like the Common Agricultural Policy (CAP) and Fisheries Policy, from which we will have complete freedom after Brexit to do what we like – except, we will still have to deal with the EU.

Here is a foodie microcosm of the sort of negotiations we face with the EU when leaving. Part of that divorce will be getting our fair share of the 'European wine cellar'. There are over 40,000 bottles of wine kept in EU buildings for various EU functions. The UK could claim its share as one of the 28 member states – i.e. around 1,400 bottles. But if I were negotiating, I'd say the number of bottles we claim should reflect the amount of money the UK put in – about an eighth of total EU contributions, meaning around 5,000 bottles. But that does not take into account the quality of the wine. I can see the French negotiators saying we should leave their Chateauneuf du Pape and take the Rioja.

The man to open negotiations, 19 June 2017, on behalf of the UK was David Davis. He spent over 15 years working for Tate & Lyle – the big sugar producer in East London. He spent much of his time battling with EU regulations that made it difficult to import sugar cane from the Caribbean – on which Tate & Lyle depends. Tate & Lyle were one of the few food companies to campaign actively for Brexit.

After leaving Tate & Lyle, David Davis wrote a management book entitled *How to Turn Round a Company.*[2] There he says: 'A general air of visible determination and activity is extremely important to the perception-shaping exercise.' No worries there then for our negotiations with the EU.

Davis spent years haggling over corporate restructuring at Tate & Lyle. Here are some more of his golden rules, which those in Brussels may have read. He says adequate preparation is vital. 'The first essential step is to view your problem from the perspective of the other side … Understanding clearly the intent of the other side is the first step to a mutually successful negotiation … The risk for everyone involved [in negotiations] is very large, and this puts a premium on nerve.'

As a trade union tutor I taught union representatives about negotiating, saying 'know what your target is, and know the red lines'. I was also involved in many negotiations at county level for the lecturers' union NATFHE (National Association for Teachers in Further and Higher Education). Compared to the EU negotiations, these were simple, but nevertheless we could spend days over a single sentence. I also taught negotiators how to read upside down – so they could read management's notes.

Opportunities

Andrea Leadsom, then Secretary of State for DEFRA, said at the NFU Conference in 2017 that we can

> ensure a more tailored approach – one that recognises the needs of hill farmers alongside those of arable farmers and protects our precious uplands as well as our productive fenland. These are the kind of questions the current system can't even pose. I am determined that we will do so much better for farmers when we leave the EU – with a system based on simpler, more effective rules, we'll be free to grow more, sell more, and export more of our Great British food.[3]

Standards

A YouGov survey for Friends of the Earth, in August 2016, asking whether people wanted to keep EU standards for the environment, claims that 83 per cent of those polled said they wanted higher or similar standards on leaving Europe. A major difficulty is that if we just move some EU laws into British law, they refer to EU regulatory agencies that aren't easily replicated. There is a possibility we have some institutions similar to the EU ones – the Expert Committee on Pesticides, for example, can be compared to the European Food Safety Agency – but on the whole nobody is talking about this, especially DEFRA. Another question is whether DEFRA has the skills and capacity to cope with adopting a barrage of potential new laws.

While the general population, well away from the fields, may want higher standards, they will be up against the National Farmers Union (NFU). In an EU consultation, the NFU said that the Directives for Birds, Habitats and Water, need to be 'more balanced' with 'food production'.

The NFU doesn't like being told what to do, and wants a more flexible approach. Members of the NFU have made clear their opposition to the EU moratorium on neonicotinoid insecticides (popularly 'neonics'), as we will see in Chapter 9. The NFU was in the EU Court in February 2017 supporting the case bought by Syngenta and Bayer to overturn the moratorium on neonicotinoid use.

The government says it will be guided by five principles in its re-interpretation of EU mechanisms. They are Trade (Export), Productivity, Sustainability, Trust and Resilience. Leadsom said: 'we have an opportunity to take a fresh look at these schemes and think about what mechanisms are needed to promote the twin goals of productive farming and environmental improvement. I want to consider, for example, how we will strike the right balance between national frameworks for support measures whilst tailoring them to local landscapes and catchments.'[4]

By 'support', she is here referring to the subsidies that the EU used to pay to landowners, but which are now in the hands of the UK government, as we will see later. However, she is also saying there is an opportunity to shake up the way the various schemes protect our food and environment. I hope this means there is an opportunity to coordinate the various aspects of food, farming and the environment with stakeholders at local and regional level to come up with a shared vision of the future. There are opportunities to make more sense of the directives, by bringing them together in an integrated whole, that takes in soil, water, air and labour.

I would especially like soil to get a look in. We need a National Plan for soils. We will see in Chapter 5 how the NFU and the 'new' Labour government blocked the EU Soil Framework Directive. A recent Environment Audit Committee says we need a Soil Plan and some monitoring.[5] The present government responded saying we don't need to monitor our soils, as they don't change very fast.

As we take up issues that have previously been controlled by the EU, the possibility of passing on responsibilities to our four member countries and the regions becomes a hot potato. Agriculture is a 'devolved' issue. It is already clear that our four nations will see their own food and farming systems differently. Three have Agricultural Wages Boards – England doesn't. There is also the matter of different sorts of land. Scotland has the highest proportion of poor land, so receives more EU funds. Northern Ireland gets 10 per cent CAP funding when per capita averaged over the UK it would be only 3 per cent. Each nation will interpret the laws coming from the EU in different ways. Scotland says it doesn't want GM, but the other countries are likely to differ.

Some of our foods have 'Protected Food' status. Famously, Melton Mowbray pies won the right to protected status, so only those pies made in Melton Mowbray in a particular way can carry the name. Another classic is Stilton cheese: it has to made in Stilton – and in a particular way. These are part of a scheme whereby the EU protects particular foods throughout the EU.[6] There are 78 Protected Foods in the UK. The EU may not 'protect' them in future when we are outside the EU, meaning companies elsewhere in the EU may make the same product and claim it as the real thing.

Subsidies

The biggest opportunity in the Brexit process is to redirect the £3bn EU CAP funding. I suggest we stop funding rich landowners and subsidise labourers. I will make the case that we should move subsidies from landowners and make the money work for us, by giving it to landworkers. We need to subsidise labour in the food sector to keep food prices down, which customers demand. This will fund local produce and rural communities. It will be worth it in the end.

By my calculations, the £3 bn could fund 300,000 UK national farmworkers and farmers for around £10,000 each annually. Instead of – a lot of – money going to the landed gentry to do nothing, this would be paid to people to work. Surely this would: help farmers pay their workers better and so reduce their overheads; encourage small farmers to pass on their farms to their children earlier; attract younger workers into the sector; help replace migrant workers with permanent workers.

Challenges

Subsidies

Farmers may have been under the impression that nothing would change post-Brexit. The then Farming Minister George Eustice did nothing to suggest otherwise. He told a farmers' meeting in Wales a few days before the Referendum that the subsidies (previously from EU) were guaranteed (as UK funds). Eustice, who campaigned to leave the EU, said they would enjoy 'as much support' as they received from the EU if Britain left the organisation.[7] He added that leaving the EU would be an opportunity to deliver change for the Welsh farming industry. This may have accounted for quite a lot of that Welsh Brexit vote. A month later, after the Referendum, the same minister said he could not 'guarantee that future agricultural support programmes will be as generous as current EU subsidies'.[8]

Eustice also suggested that instead of just paying out the CAP money, we use it as an insurance scheme, just as the US runs its subsidies. It holds money back and forks out when farmers do not get the returns they expect. It sounds like keeping the money for a rainy day. Except in the mad world of food markets, that insurance money is needed when there is a good harvest.[9] So President Trump is threatening to slash it.[10] He intends to put a $40,000 limit on crop insurance premium subsidies, which could prevent some farmers from insuring their entire acreage. There is currently no limit. 'Changes in crop insurance that cause farmers to cut back on plantings could provide relief to a global balance sheet heavy with supplies. Falling production could pay off in the long run by helping to lift prices', said Scott Irwin, agricultural economist for the University of Illinois.[11] Again, this reminds us: the problem is over-production.

Another option is doing away with subsidies altogether. The likes of the *Financial Times* say that agriculture can benefit from doing away with subsidies and 'embracing competition',[12] citing what happened in New Zealand in the 1980s. Then, New Zealand got rid of its subsidies and established trading links with countries nearby, like China, which agreed to take more of their food – especially meat. The trouble is our closest market is the very one we are leaving.

There will be other bodies eager to get their hands on the subsidies. The National Trust, claiming to be the biggest farmer in the UK, was quick to say that subsidies should only reward farmers who promote conservation.

Also lost will be the EU Cohesion Policy. This is the name behind hundreds of thousands of projects, which aim to reduce disparities between the various regions, especially the 'least-favoured regions', and to promote 'economic, social and territorial cohesion'. Scotland has a lot more 'less-favoured' land than England. The Cohesion Policy was intended to promote more balanced, more sustainable 'territorial development'. While the UK overall was not seen as one of the most disadvantaged, within the UK there are clear disparities. Quite how that will be resolved is anybody's guess. The European Regional Development

Fund (ERDF) and the European Social Fund (ESF), which invested €476m and €464m respectively between 2014–20 in the UK, will go.

Standards

We will hear a lot about cutting 'red tape'. But the bottleneck isn't EU red tape. It is DEFRA's. The EU fined them E500m for taking so long to distribute £3bn of their simple 'Basic Payment'. A recent Government Select Committee said this doesn't bode well if we want DEFRA to sort out what may be some complex funding arrangements.

There lie ahead staggeringly difficult negotiations about laws, regulations, protected food, environmental directives, food standards and health requirements relating to food. 40 per cent of all environmental directives relate to food and farming. They include nitrogen vulnerable zones that are based on an EU directive,[13] National Action Plan for Sustainable Use of Pesticides,[14] Habitats Directive, Working Time Directive,[15] OSH Framework Directive[16] and Animal Welfare. Nobody knows how these will translate when Britain splits from the EU. Even the UK definition of (food) waste could change, as that is based on the European Court of Justice.[17] These EU directives give rise to our laws on water pollution, biodiversity and habitats, and they could all be rewritten. The Environment Minister, Michael Gove, wants to scrap 'absurd' rules like the Habitats Directive.[18] One of the reasons he gave for supporting Brexit was the EU limitation on the maximum size of containers in which olive oil can be sold.[19] I'm glad his priorities are clear.

DEFRA will have to deal with some 1,200 pieces of EU legislation. Somebody has to go through all these pieces of law and decide what can be 'lifted and shifted' into the proposed Repeal Bill, and identify the other parts which are 'inoperable', i.e. too closely linked to the EU. We will see much use of something called the Henry Eighth clause, which allows legal changes without parliamentary scrutiny.[20] And this is at a time when DEFRA is undergoing a massive reorganisation, to consolidate the work of 33 separate agencies. I have seen some jobs being advertised, but its annual budget is scheduled to be reduced by 35 per cent by 2019.

There are also standards that the EU adopts as part of its global commitments, like the Convention on Biodiversity. These will still apply, although the Precautionary Principle part of the Rio Convention (1992) may get revisited, as many in the UK consider the EU's interpretation of the Principle is misguided, as we will see in Chapter 10.

We will have to replicate the EU Standards Agencies. We have agencies that are capable of replacing these agencies – at a cost. We could set different standards, say, in animal welfare or pesticides. But I can't see the point, if we want to export into the EU – our biggest market – as they would say the produce is not acceptable. The other downside is that in the past, members on our committees would take time out to sit on the European Standards bodies and try to influence them. Now we won't be able to.

Scientists were overall very Remain. They could see that if we want to be at our best, we have to work with others, and there were a lot of big EU-coordinated projects. It is not just about money. EU-funded courses and programmes like Erasmus encourage scientists to mix across the continent. The EU wants programmes that enable us to find out about each other's cultures and ways of doing things. Joint EU collaboration with the UK (in 2017) worked out how the peach aphid spreads viruses on crops.[21] Science will be involved with many of the changes brought about by Brexit – particularly with the approval processes for pesticides and genetically modified crops. There will be increased scrutiny of the role of scientists in these areas, so there is a whole section devoted to this topic – Part III – in this book.

DOING THE DEAL

We were told we will get the best deal. But it is hard to see what that means, with so many competing interests. We explore here what 'doing a deal' might mean.

Out of the Single Market

The Prime Minister considered that if the drive in Brexit was 'immigration' and 'control over our borders', we would have to come out of the Single Market. Many 'soft' Brexiteers wanted control over migration, but also open access to the Single Market – the cake and eat scenario. The Single Market requires the four 'freedoms' – 'freedom of movement of goods, capital, services and people'. So if we want control over EU immigration then we have to leave the Single Market.

The Labour Party says it wants Brexit but also wants free access to the Single Market. The Single Market depends on the free movement of labour, at whatever skill level. The general interpretation is that if we want control over immigration we must leave the Single Market.

The Single Market allows the easy movement of food goods across the EU without any border controls. There are no checks on the health, safety or origin of foodstuffs, because 'non-tariff' barriers have been removed. This is not just a matter of final food products crossing borders, but allows longer food chains. For example, Guinness sends its famous stout to Belfast to be bottled and then returned to Dublin. Most Irish Cheddar cheese is made from Northern Irish milk, with no border checks.

The Single Market removes 'non-tariff' barriers, so the same rules apply for packaging, hygiene and safety standards across the EU. We will face these 'non-tariff' barriers when we leave the Single Market and want to get back in. The lack of non-tariff barriers makes the Single Market more than a free-trade zone.

But Inside the Free-trade Zone

A free-trade area is one where there are no tariffs or taxes or quotas on goods and/or services from one country entering another. There is a free trade zone in Europe and we helped to create it: the European Free Trade Association (EFTA), which counts Norway, Iceland, Switzerland and Liechtenstein as members. There are no tax barriers to movement, but the boring rules to harmonise non-tariff (e.g. food safety) barriers are missing, so checks would have to be made crossing into the EU.

Norway sets its own tariffs on goods imported from outside the Single Market. Norwegian goods (with exceptions for farm produce and fish – because of the subsidies!) are imported tariff free into the EU. Switzerland has negotiated a series of bilateral deals that give it access to the Single Market for most industries, although it also has to apply EU rules and pay the EU money. This deal is under threat of renegotiation after a slim majority voted in favour of imposing limits on immigration in a 2014 Swiss referendum. We know the EU insists on free access for EU citizens.[22]

So while there may be some advantages to the 'Norway model' – i.e. no tariffs going into Europe – that isn't much use for the food and farm sectors, as they are excluded from this arrangement.

Outside the Single Market but Inside the Customs Union

One country – Turkey – enjoys this arrangement. It is basically a bilateral trade deal. 'The Customs Union entered into force on 31 December 1995.

It covers all industrial goods but does not address agriculture (except processed agricultural products), services or public procurement. Bilateral trade concessions apply to agricultural as well as coal and steel products.'[23]

Leaving the Customs Union Too

The European Economic Community was a Customs Union before the EU. So the EU is both a Single Market (common internal non-tariffs) and a Customs Union (common external tariffs). If we also leave the Customs Union, we leave all the tariff protections too. If we come out of the Customs Union, we will face EU tariffs when moving food into the EU.

However, there has been confusion as to whether we are coming out of the Customs Union as well as the Single Market. Up until 25 April 2017, the Labour Party seemed as unclear as everybody else. Then Sir Keir Starmer, the Labour Brexit spokesperson, criticised the government for taking the idea of continued membership of the Customs Union 'off the table'.[24] After the June 2017 election, differences emerged within the Labour Party. Jeremy Corbyn and John McDonnell say the party is committed to taking Britain out of both the Single Market and the Customs Union, while the shadow Brexit and shadow trade secretaries say the UK should try to negotiate a new form of Single Market membership. There are similarly quite different views in the Conservative Party, and many tensions are likely to emerge. Europe has been known to tear the Conservative Party apart. All variations on these themes will no doubt be played out over the next few years. That is one of the purposes of this book – to help you decide what we should go for.

Coming out of the Customs Union means we can do our own trade deals, across the world, and set our own tariffs. But it also means we would face the tariffs that the EU imposes on other countries. By coming out of the Customs Union, lorries become subject to customs checks to collect taxes. Many 'free traders' want to 'do deals' with the rest of the world. In order to do so, we would have to leave the Customs Union.

Brexiteers v Free Traders

I have made a distinction between those who want 'control over our borders' – coming out of the Single Market – whom I'll call 'Brexiteers' –

and those who also want to step outside the Customs Union – the 'Free Traders'. Coming out of the Customs Union is as big a deal as coming out of the Single Market.

We could come out of the Single Market and make a bilateral deal with the EU – and not come out of the Customs Union. When we hear talk of 'doing a deal', I wonder what is meant. Brexiteers could mean coming out of the Single Market (and getting immigration control), but remaining in the Customs Union.

The 'free traders', on the other hand, want to come out of both the Single Market and the Customs Union. In which case, they have 'to do a deal' on all the 2,000 tariffs involved, before setting off round the world. This is a staggeringly complex operation, as agricultural tariffs are considered highly sensitive.

There are profound implications about what our status would be outside the Customs Union. We would have to make arrangements with the EU as a new trading partner and special arrangements with the WTO. The WTO says we would have to 'start from scratch'. Thousands of agricultural products would have to be renegotiated with countries around the world. It is monumentally complex, making Google algorithms look like abacuses, as we will see in Chapter 4.

The EU will prefer to see an outline of the divorce settlement before discussing trade deals. We will need teams of negotiators, but it appears that most of our negotiators work either for the EU or the WTO. To leave both the Single Market and the Customs Union is a monumental decision. I bet not one person in a hundred who voted in the Referendum understood this distinction.

The free traders are doing what British capital loves to do in a crisis – running away. Capital would rather invest anywhere in the world where it can find cheap labour, rather than invest in our own infrastructure. With the prospect of fewer migrant workers in this country – because we are controlling our borders – capital will look elsewhere. It is hard to make money in rural areas, especially the higher up the hills you go.

We have known this for 150 years. You get lower returns on investment in poor land than you would expect if you were to invest in better land. This is blindingly obvious, but usually ignored. I heard a farm consultant say farmers are now getting more precise about where they put their money. Even within a farm they will invest in a GPS tracker system or drones, in particular fields, often not bothering on the rougher bits. He added that the top 5 per cent of producers were not bothered too much about the loss of subsidies, but wanted to 'compete on a level playing field'. These producers will have the loudest voice, and will be heard all the way to parliament.

The trouble is we are talking not about 'playing' fields, but 'ploughing' fields. There are no level ploughing fields in the Pennines near where I live. It is a lot trickier to plough hills here than it is with two miles of level fields using GPS controls. This uneven investment goes on big scale. According to Rural Business Research,[25] the income for an average arable farm in the East is about £74,000, £40,000 of which comes from agricultural support. The income for farms in 'less-favoured' areas averages around £17,750, of which £17,100 comes from support! In 2016, the government's estimate for Total Income From Farming (TIFF) showed average incomes for farmers at just over £20,000, with a total of £3.6bn, representing a disastrous fall of 7.5 per cent for the whole of the UK. No wonder farmers need EU subsidies, now accounting for 80 per cent of farm income compared to 50 per cent just three years ago.

All our efforts and investments are heading abroad, while our rural areas will find it harder and harder to attract investment. Northern Ireland and Scotland and Wales will feel the brunt of this. Look at Scotland – 85 per cent of its land is considered 'less favoured', compared with only 15 per cent of land in England. Quite simply, our green and pleasant land will suffer, if the hunt for cheaper food takes off across the world.

Tariffs

When we come out of the Customs Union we will realise what a Tariff Cliff is. The EU is surrounded by food tariffs of between 5 per cent and 600 per cent, designed to protect EU food producers and farmers. Consumers in the UK may get cheaper Thai prawns, but UK farm and food producers will be climbing the Cliff.

It is likely that some trade deals will be done to maintain some free access to the Single Market, like the car and aerospace industries. But the food and farm sectors have much smaller pieces of the cake, and could be overlooked. Added to which many Free Trade Agreements (FTAs) have agreements on reducing tariffs – except for 'sensitive products' – most of which are agricultural.[26] For foodstuffs, especially meat, we will be looking up at a very high Tariff Cliff to get on to the table top of the EU Single Market, where we export most of our meat.

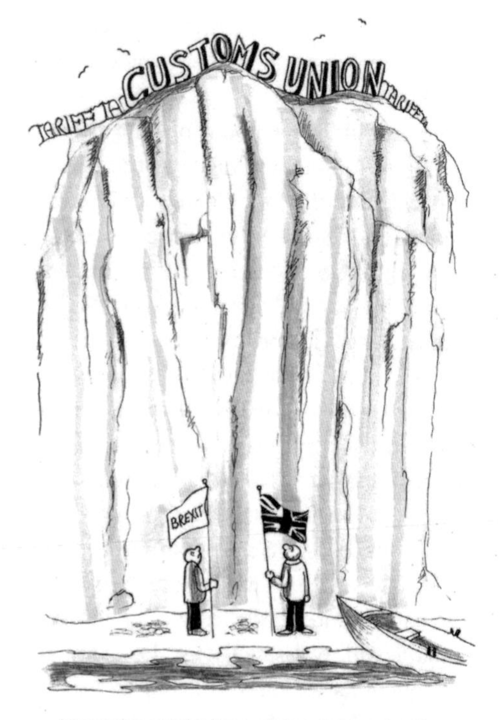

I thought you said the view would be better from down here

If we examine the regimes of specific tariffs for important foodstuffs, we will find that it is not just our farmers who are protected in the EU, but many roasters, grinders and processers of food. Raw foods – like coffee beans – are not usually taxed. But processed versions – like roasted coffee beans – are taxed. This is a major disincentive to prime producers to develop their own technologies to 'add value' to the raw foodstuff. We add value here.

Most Favoured Nation (MFN) is a level of status given to one country by another and enforced by the WTO. Countries with MFN status are given specific trade advantages. It means that when the UK leaves the EU it cannot be taxed more than the MFN. The EU cannot discriminate against us, nor can we against them.

Outside the tariff wall, Table 1 shows the tariffs that UK exporters into Europe would face. Instead of these tariffs protecting our food and farms, these same tariffs will become the barrier we will face when entering the EU. And when we turn around the other way, we see across the seas, our old friends rubbing their hands at the prospect of flogging us more corn and grain, because now they won't face tariffs. As Table 1 shows, the EU slaps a 30 per cent tariff on dairy and many meat products. That's a powerful disincentive. In future, it could also make it difficult for us to export dairy products to the EU.

Table 1 Tariffs between the EU and MFNs (2013)

	Average %
WTO agricultural products	14.8
Animal products	20.4
Dairy products	31.7
Fruit, vegetables and plants	13.3
Coffee, tea, cocoa and preparations	11.6
Cereals and preparations	18.1
Oilseeds, fats, oils and their products	7.5
Sugars and confectionery	25.4
Beverages, spirits and tobacco	14.2

Source: WTO EU Trade Policy Review, 2013.[27]

Just to make things even more complex, these tariffs are all linked to 'quotas'. These are the levels before taxes kick in. They would be up for negotiation too. For instance, sheepmeat (lamb and mutton) attracts a tariff of around 40 per cent – depending on which bits of the sheep are being sold. However, New Zealand has a quota to export over 200,000 tonnes to Europe, around 65,000 tonnes of which comes here, without tax. Australia is allowed 20,000 tonnes of quota. Quite how the New Zealand quota would translate post-Brexit is tricky. It could be 200,00 divided by 28, the number of EU member states, totalling around 7,000 tonnes, or we could use the present amount coming to the UK as our own quota. This trivial-looking debate about numbers makes a massive difference to sheep farmers here.

Immediately after the Brexit Referendum, people started talking about 'free trade', playing by 'WTO Rules'. When Theresa May said 'no deal is better than a bad deal', she was referring to coming out of the EU entirely and our trade being determined only by the rules of the World Trade Organisation (WTO). This is what those rules mean.

The WTO is a club. If you join, you have to follow the rules. The WTO prefers to encourage negotiation but can instigate a dispute panel, and just occasionally has sanctioned members.[28] At present the EU acts on behalf of its 28 member states in any dispute within the WTO Rules. Cases taken against the EU have been as varied as Canada challenging duties on cereal, Chile disliking the EU description of scallops and Brazil complaining about importation of poultry products. The USA tried to do away with 'the Protection of Trademarks and Geographical Indications (GIs) for Agricultural Products and Foodstuffs'. As an individual member we would have to defend ourselves against any such challenges.[29]

Once we have divorced from Europe, and with an EU trade deal still pending, the UK would seek to trade on the basis of WTO rules. But the WTO's chief, Roberto Azevêdo, said that because the UK joined the organisation as part of the EU, we would have to renegotiate the terms of our membership as a fully independent member. The UK will have to detach itself from the EU, then regularise its position within the WTO before doing any deals with other WTO members. There are no precedents for this and it could – probably will – take a very long time.[30]

The EU has become a net exporter of food and drink, with EU agri-food annual exports estimated at a total of $122bn. The EU has increased these exports over 8 per cent over the last ten years, mostly in primary and processed food preparations. The UK will be in direct competition with this volume of trade when trying to increase UK exports.

Opportunities

It is hard to see any real opportunities in Europe for boosting our food exports.

Challenges

Once outside the EU, instead of the Tariff Cliff protecting our food and farming, we will be facing it. At the border, customs officials will take much longer to check cargoes – for tax purposes. There will be no EU funding to encourage selling our food into Europe anymore. Most of the farming community keeps saying they want 'access' to the Single Market: the 'cake and eat it' approach we've seen so much of. We know we can't cherry pick, so that doesn't leave much to chew on.

Meat is particularly affected. We export 40 per cent of the sheepmeat we produce, and 96 per cent of that goes to Europe. 80 per cent of beef exports and 60 per cent of pork exports go to Europe. Welsh farmers received £3m from the EU to promote red meat to niche markets in Germany. They won't get that anymore. Meat exports will be especially hard hit. Not only will tariff-free meat come in from all over the world, but we will face 30–40 per cent tariffs if we try to get ours into the EU. I can't see the EU agreeing to quotas for our sheepmeat.

Sanitary Measures

Sanitary and Phytosanitary (SPS) sound like something to do with hygiene. They are health, safety and welfare requirements that countries can impose on food. They can have a greater effect on trade than tariffs for some countries. However, they must reflect standards already existing in the country.

The food industry is international. A food product may be produced in one country, processed in another, and sold in a third. The EU has always put a premium on consumer safety, leaving it to member states to use their own methods to provide this. There are EU labelling regulations, requiring notice of allergens, and nutrition information,[31] which is the responsibility of the Food Standards Agency, who leave enforcement to local councils and their public health officials. Whether this will be enough to provide reassurance in the future is open to question. If we want to track adulteration or mislabelling, cross-border regulatory cooperation is critical.

Currently, the UK's food quality and regulation is part of an inter-dependent network of agencies working together across the EU. After Brexit, the UK is removed from this network, so the costs will have to be borne by the UK alone, and are likely to be greater in the future. Our UK food regulators will be isolated from their European counterparts and their intelligence. It is hard to see how our food integrity will be better protected.

SPS measures spell out standards on imported food,[32] and should be based on the WTO SPS Agreement, international standards, recommendations or guidelines or on scientific principles. However, countries often impose unjustified SPS measures, which can affect other countries' exports of agriculture and fishery products.

The WTO says:

> All countries maintain measures to ensure that food is safe for consumers, and to prevent the spread of pests or diseases among animals and plants. These sanitary and phytosanitary measures can take many forms, such as requiring products to come from a disease-free area,

> inspection of products, specific treatment or processing of products, setting of allowable maximum levels of pesticide residues or permitted use of only certain additives in food.
>
> The basic aim of the WTO SPS Agreement is to maintain the sovereign right of any government to provide the level of health protection it deems appropriate, but to ensure that these sovereign rights are not misused for protectionist purposes and do not result in unnecessary barriers to international trade.[33]

The EU insists on some agricultural products having strict standards, including GMOs, the way beef is produced and chicken is treated with disinfectants. These would be included in any US-UK trade deal, and the US would probably challenge them on scientific grounds.

Environmental standards are even harder to justify. China finds some of these standards a serious barrier to EU trade. It may be tempting for the UK to drop some of the EU standards – and attract more trade from, say, China. But we may then lose out on trade with the EU. Or we could 'up' our standards to find new niche markets. Whichever option is chosen, we will need scientific evidence to justify our decisions.

Challenges

I find it hard to work out how there can be a 'deal for all'. When Theresa May fired the starting gun for Brexit she promised she would get 'the right deal for all',[34] but it is hard to see how one deal can fit all. The basic contradiction is that by removing tariffs on food imports, consumers may get a better deal, but producers will get a worse one. This applies to a whole range of foodstuffs, like dairy, cakes, biscuits, meat and fish.

The next few years will be decisive. I would hope we will be able to have an open debate about what we want. When you look at the complexities for food alone, you can't help but think that if negotiating a Brexit with 27 other countries could take a long time, any new trade deals will take much longer.

PM May said at the Conservative Party Conference in October 2016 that Brexit is 'a turning point for our country. It is a once-in-a-generation chance to change the direction of our nation for good, to step back and ask ourselves what kind of country we want to be.' I'm going to paraphrase that and say we have 'a once-in-a-generation chance to change the direction of our nation that would be good for food'.

3
Moving On

In this chapter we look at what happens after Brexit – whatever happens. The harder the Brexit, the more we will be on our own. There are two main directions of travel – 'going global' or 'buying British'. The free traders want to go global, while buying British looks to ourselves to produce better food in a way that will also help our economy.

GOING GLOBAL

We hear a lot of talk about going global. At present we cannot – because we are part of the EU Customs Union. You may be under the impression that there is a world of free trade out there, governed only by WTO rules. That couldn't be further from the reality, as we have seen. But that is what the free traders keep promising.

Just about every country has its own tariffs and quotas, particularly for foodstuffs, based on all sorts of historic niceties. The purpose of free-trade deals is to avoid these tariffs and quotas. The norm is tariffs – to avoid them we need Free Trade Agreements (FTAs). FTA agreements

Ok, but apart from the 3 billion euro subsidy, protection from non-European suppliers, seamless movement of food across borders, adapting other countries to our animal welfare standards and selling our red meat to Germany and our lamb to France - what has the EU ever done for us?

on just one food stuff can take many years. And it is hard to get a deal on just one foodstuff. The other side will want concessions in return – that is what is meant by a deal. For example, the US would require in any trade deal (on electronic goods/ aerospace equipment) to include agricultural products – and that means tax-free entry for their cheap hormone-drenched beef, pork fed with ractopamine (see Favourite Foods: Pork), chlorinated chicken and GM corn.[1] I'm more worried about the corn itself, rather than the GMO.

The word 'globalisation' is repeated ad nauseum. We were told at the Labour Conference 2005, by Tony Blair that:

> What we can't do is pretend it is not happening. I hear people say we have to stop and debate globalisation. You might as well debate whether autumn should follow summer. They're not debating it in China and India. They are seizing its possibilities, in a way that will transform their lives and ours. Yes, both nations still have millions living in poverty. But they are on the move. Or look at Vietnam or Thailand. Then wait for the South Americans, and in time, with our help, the Africans. All these nations have labour costs a fraction of ours.[2]

It is interesting he picked out cheap labour, ever present in our food chains, as we will see in Chapter 5. He was right in that China is embracing 'globalisation'. The Chinese President, Xi Jinping, at the Davos summit in 2017, extolled the virtues of globalisation. China joined the WTO in 2001, but only after it had to agree to cap its agricultural subsidies to 8.5 per cent of total output. They wanted 10 per cent, but the US wanted less – to protect its own farmers.[3]

Globalisation is now being used to explain why people voted for Brexit and Trump. This 'populism' is said to appeal to the many people who feel they have 'been left behind by globalisation'. It is curious that it is the two nations who go on most about 'free trade' that are the ones most affected by 'populism'. It is a warning that the movement of capital round the globe searching out new markets forgets the people it left behind – those not working in the mills, but those labouring in the fields, and those working in dull food-sector jobs.

These two nations are taking opposite paths to deal with this issue. President Trump trumpets that he promises to invest $1 trillion in American infrastructure – rather than overseas. This is usually called 'protectionism' – look after your own first. Globalists say this is not the way to make money. The Conservative government, on the other hand, wants to recapture some old sense of Empire and go global, which the German newspaper *Die Welt* called 'Little Britain'. The signs, surrounding the Prime Minister when she announced we are coming out of the Single Market at Lancaster House in January 2017, said 'A Global Britain'.

The WTO have various bodies that act internationally to set food standards, which we would use. The most important is called *CODEX Alimentarius*, which establishes the international harmonisation of food standards. Importantly, the SPS Agreement (the sanitary measures that are part of trade) cites *CODEX*'s food safety standards, guidelines and recommendations for enabling international trade and protecting public health. It is a science-based organisation of specialists in a wide range of disciplines. *CODEX* has developed over 200 standards covering processed, semi-processed or unprocessed foods, plus 40 hygienic and technological codes of practice, evaluated over 1,000 food additives and 54 veterinary drugs and set over 3,000 maximum levels for pesticide residues. If we come out of the EU, we would probably use these standards for trade. We will not be able to change or influence them, as the UK does not have a place on the *CODEX* committee. The EU does.

Opportunities

Exports: The then Environment Secretary Andrea Leadsom announced an 'ambitious new action plan' in Paris to sell more exports. The government claims[4] global ambitions to export more food and drink could bring in £3bn. The 'International Action Plan for Food and Drink'[5] identified nine markets across 18 countries with a hunger, or rather thirst, for classic British items. These include an extra £185m in exports to Japan for tea, jam and biscuits, nearly £300m to Australia and New Zealand for our beer and cider, and over £200m in Mexico and Latin America for our whisky and gin. They aim to get an extra £400m for pork, beef, lamb (the poorer cuts) and poultry, as well premium seafood. They are targeting drinks for China as they are the second largest imported beer market in the world. Also, 'The regional hub for the wider Middle East and North Africa and a strong ex-pat market provide opportunities for premium grocery, an emerging organic and "free-from" market, as well as for confectionery, drinks and snacks, that amounts to £150m over

5 years.' In total these new exports add up to £3bn, about a tenth of the Food Trade Deficit.

Imports: If we are linking up with old friends, then Kenya is the place to go. When Kenya gained its independence from Britain, in the mid 1970s, the British left most of the land to the Kikuyu people and their leader Jomo Kenyatta became President. His son is now President. I predict a deal will be made with Kenyan exporters to bring in more fruit and veg, but we are walking into some history – as we'll see in Chapter 4 – and environmental issues – as we'll see in Chapter 6.

We should see the Commonwealth not as a place for trade deals, but for co-development. We love our chocolate and have a good history encouraging small cocoa producers in West Africa. But much of their crop is diseased and lacks investment to replace. They will soon be in competition with South American cocoa plantations. We want a direct relationship that feeds our insatiable appetite for chocolate and also looks after cocoa farmers and the workers' future (see Chapter 11).

Apparently Whitehall officials describe plans for Britain's post-Brexit trading relationship with the Commonwealth as Empire 2.0, conjuring up software connotations. This national conversation about the lost empire is all too often focused on what 'we' did or did not do. That is part of the reason I have used 'we' throughout this book. 'They' – the peoples of the former empire and the current Commonwealth – do not feel the same. When I was working in many Commonwealth countries, it wasn't long before the conversation turned to how they dealt with Britain's former presence. Opinions differed, but most Commonwealth countries are still living with an uncomfortable legacy from the old colonial past, which they do not see in the same way we do.

Challenges

If the UK gives up free access to the Single Market, and comes out of the Customs Union (hard Brexit), while trying to negotiate our way back into parts of it, the UK may simultaneously want to set up food trade talks elsewhere. But many governments will prefer to wait for the outcome of Britain's talks with the EU. For instance, it matters greatly to (other) third countries whether or not the UK joins the European Economic Area (EEA), and stays in EFTA, as we saw in Chapter 2.

I do wonder how much former colonies can deliver. I have worked in 20 Commonwealth countries for various trade unions, talking a lot with workers about our food practices. Most of the people I talked to were aware of the vast distances involved. Apart from cocoa and coffee, and vegetables in Kenya, there isn't a lot of potential to increase food imports and exports with Commonwealth countries.

We import a lot of food from already 'water stressed' countries. These are countries like South Africa, Egypt, India and Spain where water outputs are greater than inputs. We will see in Chapter 7 how much 'virtual water' is used to grow our foodstuffs. That puts pressure on water resources in African countries, making it harder for many people living there to access clean water for their own use. There is *actual* water in the fruit and veg flown to our rain-drenched shores.

Then there are some old enemies around. In November 2015, a UK government delegation wore Remembrance poppies on a visit to China. However China remembers the opium wars when we grew 100,000 acres of opium to drug the Chinese to get money to buy their tea to satisfy our own addiction. When the Chinese sank the opium, we sent in the Navy and forced them to give us Hong Kong and open their ports to 'free trade'. They asked our delegation not to wear the poppies, but we refused. China now has a presence in just about every country we may want to trade with – in Asia, the Pacific, Africa and South America. I suggest that when we meet up with Chinese investors we avoid Remembrance Day.

The main large countries we can trade with are known as the BRIC countries – those developing countries that have increased their share in global markets dramatically in the last 25 years. They are Brazil, Russia and India as well as China. We traded food with Russia until recently, but they have nothing to do with us now, after the imposition of EU sanctions against Russia following their interventions in the Ukraine. Russia retaliated overnight by banning the import of any foods from Europe.

The UK is the second-biggest importer of Brazilian beef, accounting for about 20 per cent of sales to Europe. Brazil believes that having direct access to the UK may work in favour of Brazil's beef exports, as their shipments to the EU are affected by trade restrictions. Fernando Sampaio, executive director of the beef export organisation Abiec, said: 'I believe the UK will be more liberal compared to the EU, but it's impossible to say how exports will behave because we still don't know how access market rules will change.'[6] They see import potential in the UK – provided we relax the standards imposed by the EU (see Chapter 4).

Rule Brexitannia?

BUYING BRITISH

As part of our departure from Europe, we will have to try to sort out who we are – our identity – and we will look at that more in the final chapter. The Prime Minister asked us to take a step back and ask: 'what kind of country do we want to be?' However, she should have said 'countries'.

The four countries that make up the UK differ in their approaches to food and farming. Farming is already a devolved matter, but more tensions will appear. Some may want to distribute subsidies in different ways. We cannot change the geography of the countryside, but we can change what we do with it.

Northern Ireland has more mixed farms, and a lot of foreign workers in its major food preparation plants. It will now have a border with Ireland, with whom it trades a substantial volume of milk. Wales is primarily composed of family farms, and while they voted Brexit they will now be concerned about the farm support subsidies that were promised but are now going by 2020.[7] There are vast differences between farming in Western England, where family farms on pasture dominate, and Eastern England, where arable monocultures dominate.

Andrea Leadsom, the former Environment Minister said: 'The British people have handed our food, farming and fishing industries an extraordinary opportunity to thrive.' They have indeed. We will need coordination from the government and the other big players – particularly the retailers. In recent years several retailers have tried to be seen as more responsible when it comes to food provision. However, it may require more than voluntary action, as all the signs are of a 'race to the bottom', as 'Industry overcapacity, consumer price sensitivity, the grocery price wars, and generalised retailer pressure on manufacturer economics, are all forces contributing to a material deterioration in the performance'[8] of the top European-focused food groups.

There are concerns as food retailers are more consolidated in the UK (the top 4 share 76 per cent) than the rest of the EU (France's top 4 account for 67 per cent) or the US (top 4 market share 36 per cent), giving them enormous power over their supply chains. A Groceries Code Adjudicator was created to stop any bullying, but with little evidence of success.[9]

There are moves to 'bring more British food to our plates', but we need some government initiative. We need to look at how to reduce the $66bn worth of foodstuff, animal and plant products that we currently import. However, retailers in such cut-throat markets will buy the cheapest, from wherever.

Buying British produce is attractive to shoppers who think they are doing their bit. But it conceals working conditions to which shoppers turn a blind eye. Yet people living near those areas of cultivation don't like the way that food is produced. They may well have expressed their distaste at 'immigration', but it was also a protest at the way we produce that food.

So I am suggesting a new branding of Britishness, a red and green version that would include decent working conditions for all in the food chain and food production and doesn't cost the earth. We will see later in the book how current food production uses more water, pollutes more water, uses too much energy, too many finite resources and clearly isn't good for our health.

I would love the UK to lead the way in showing how this can be done. Rather than ripping off other people's resources – in terms of land and labour – we could develop ways of producing healthy, sustainable food to export to the world. We should be able to grow greener, healthier food and build our food business, already the largest manufacturing sector in the UK, into a major export business.

We can treat people and the planet better. But, the problem is cheap food and we're not going to persuade shoppers any time soon to pay more. Every survey says this is the customer's first priority. All the ethical considerations – the moral consequences of our choices on other people and animals – are second to price. I too confess that I like three labels together – 'Organic', 'Fairtrade' *and* 'Reduced'. All politicians want cheap food. Gordon Brown worried about increased food prices, as did President Nixon, who told his officials 'to get food prices down for next election'. Can we square this desire for cheap food with providing decent food?

As we come out of the EU, in whatever fashion, one issue is clear. The British government will get its hands on what were the EU CAP subsidies. We have a chance to do something with that £3bn, of making it work for us. If we do not come out of the Single Market, we could still change the way the EU distributes that money.

By subsidising workers, we can pay decent wages, so we won't be dependent on cheap – mainly migrant – labour. It would keep food prices down, and in the process we could encourage local producers to feed the cities, building better rural-urban food relations. That could all help bring food imports down, so we pay ourselves more rather than pay others (see Chapters 5 and 6).

As part of our new-found Britishness, we want to bring back some old values – in particular those working-class links to nature, allotments being our prize example. Put good old values in new settings. We want to work together to produce a 'back garden of our food produce'. We also want to create some new values – for our land and our health. These are all quite feasible if we stop praying to the god of free markets.

Here are four reasons for buying British, based on the 2017 report, *British Food – Role of Food Producers*, by Professor Tim Benton, the UK's former Food Security Champion.[10]

1. Reliance on riskier trade is not good for resilience. We are leaving the largest internal market. We cannot rely on over a quarter of our food coming from the EU in the future. The wilder trade winds blowing between three powerful countries, America, Russia and China, have their own agendas, nothing to do with us. Then there is global warming and possible effects on food production.
2. Consumers have the appetite. Two-thirds of us express a preference for buying British. But we prefer not to see the conditions in which

that produce is grown. We have witnessed the consequence of producing our food in monocultures, relying on migrant workers – it's called Brexit. If this situation doesn't improve there will be further complaints.

We are a rich market. We spend relatively little of what we earn on food. If only we could be tempted to spend a bit more. We need something like a 'British Fair Trade' label. There are a lot of 'brands' promoting various positive aspects of food. The Red Tractor scheme gives the impression of 'British' but means only 'British Standards'. We tried in Unite, through the Health and Safety Executive, to get the Red Tractor scheme to include health and safety in its standard. Even the HSE Agricultural Rep was surprised by the voracity of the refusal.

3. Supporting the local economy brings benefits all round. The NFU say that for every £1 spent in agriculture there is a £7.40 return to the economy. Research in Todmorden, the originator of Incredible Edible, found that the money spent on local food went round local shops 8 or 9 times.

   Where I live, in the Ribble Valley (where 'nobody farts', according to Jay Rayner, the restaurant critic of the *Observer*), bars, cafes and restaurants promote some local food sourcing. You get served by local lads and lasses, who often know where the food has come from. Parents want them to learn to work for wages, so we pay a bit extra for the food because we know we are investing in our future. We should be employing students to pick the harvests, as many people used to do. 'The roots of the (education) scheduling rest in our agricultural heritage, when families required their children's labour in the summer to pick fruit and farm the land.'[11]

4. Grow our diversity: The one biodiversity metric that gets me most annoyed is to do with apples. There are only ever about three varieties in the shops, when we can grow 2,000 varieties in the UK. UK orchards take up around only 50,000 acres, compared to about 230,000 acres 50 years go. EU funds paid to 'grub' them up – the correct term for digging up fruit trees.

I remember the Director of East Malling Fruit Research Station telling me (while he was telling me off for having 'sideburns') that he didn't develop 'Golden Delicious' because 'the British housewife will never go for that French variety'. How wrong he was, but we can't blame him for the demise of British varieties. There are local attempts to revive British varieties,[12] but we see little evidence in the supermarkets.

We grow a lot of maize and oilseed rape, but we should be growing a much wider diversity of crops. We should be saving our local seed varieties – I'm helping some schools in Scotland to do this.[13] Production of French and runner beans has halved. The volume and the variety of cauliflower, broccoli, Brussels sprouts, peas, parsnips, cabbage, lettuce, tomatoes, cucumber, rhubarb and pears grown in the UK has also decreased over last 25 years.

Part of the problem is not having machinery that can deal with different crops. It is expensive. We depend too much on American machinery – primarily built for size. We want diversity of machinery as well as diversity of plants and animals. The Agricultural Engineering Research Station at Silsoe, where that work could have been carried out, has closed, along with three-quarters of all UK land- based research stations in the last 25 years (see Chapter 7).

Opportunities

These are enormous. While retailers seem to be making an effort, or at least making a noise about buying British food, the same cannot be said of our major food manufacturers. There are already signs that some manufacturers are turning to British eggs, but they could also use British corn to make cornflakes. Research is underway at the Vegetable Genebank at Warwick University to produce beans to replace beans imported from the US – we could call them BritzBeanz.[14]

This is the big opportunity to grow much more of the stuff we can grow well, from apples, pears, hops and beet, cattle and sheep, but most especially vegetables, where we spend five times more on imports than we sell in exports.

There is a worldwide network trying to grow more food in cities.[15] In this country, from Totnes to Todmorden, there are community schemes to grow more locally. The Sustainable Food Cities network now has about 50 members. I'm on the board of the Incredible Edible Farm in Lumbutts, Todmorden, and it is wonderful to see how a dreadful bit of land can be made into something imaginative and productive. It was awarded the best farm in the North of England in 2017 by the Countryside Alliance – yes – the 'keep the hunt' lot.

As the food chains would be shorter, hopefully there would be less processing. We need processing for chutneys and pickles and bacon. We don't need to process all the goodness out of food, so there are no nutrients left. I remember buying a Tamworth pig in the Borders from a farm that had won the breed prize for many years at the Scottish Agricultural Show. So I asked the farmer his secret. He said: 'I'm not going to tell you, but the clue is that there is an exceedingly good cake factory down the road.' He then offered me lunch – tinned soup and white bread. I realised his pigs were getting all the goodness – the waste from the cake factory – and we were eating what was left – the rubbish.

If the food chains are shorter, we can see how much of the slice of the cake we each take. Food growers and producers need proper rewards. We should develop links between our rural areas and our urban lives. This could create a new sort of economy that benefits both urban and rural communities. More people now live in cities than in the country. Our cities – not Mexico – should be the primary market for local rural producers. When I talked with a farmer on the Fylde Coast, who grows millions of brassicas, he said that in the past his cabbages went to the two big cities nearby – Liverpool and Manchester. Now, because retail chains have moved in, he has to cart his cabbages right across the country to Yorkshire. Bonkers.

We could make the case for subsidising healthy food. After all, we subsidise our health service. It may turn out cheaper to eat healthier food. I've always thought that most of the foodstuffs coming from farms are pretty healthy. They are full of fibre and valuable nutrients. It is the processing that turns it into junk – whether animal or vegetable. Gordon Brown's Food 2030 Report (2010[16]) reckoned: 'Diet-related chronic disease is estimated to cost the NHS £7bn a year... [it is] estimated that the NHS will be spending almost £20bn a year to prevent 70,000 premature deaths a year.' We could save the NHS a lot of money by providing local food that you need to chop and chew – see Chapter 8.

Challenges

Our own internal food market could be worth tens of billions of pounds. Instead of a saturated market, encouraging cheap processed foods, we could have a vibrant, diverse, healthy food market, feeding money back into our economy.

A start is being made by one of the major retailers. Morrisons are looking for the next generation of British food producers to serve their 12 million customers and they sponsored Tim Benton's report (above).[17] They want the best local producers to supply nearly 500 stores, sourcing more local food, and reducing the distance that food travels – the 'food miles' – the term created by my friend Tim Lang.

Many retailers already claim they are buying more British food. But a 'Lamb Shelfwatch' carried out in Scotland by NFU associates in Easter 2017 produced findings that were a slap in the face for Scottish farmers.[18] NFU Scotland's secret shoppers found fewer opportunities for Scottish shoppers to buy Scottish lamb when compared to 2016. The Shelfwatch results showed that, despite previous commitments by some supermarkets, Tesco, Asda, Sainsbury's, the Coop, Morrisons (despite the claims above!), Marks and Spencer and Lidl all chose to import huge volumes of New Zealand and Australian lamb, shunning home producers. In one Tesco store, underneath a banner proclaiming 'The Best Scottish Lamb in Season', a secret shopper found 100 packs of New Zealand lamb. Complaints to the Public Health Inspector failed to rectify the situation – because there isn't a law against it. Food service providers will buy frozen New Zealand and Australian lamb in preference to home-grown, because it is cheaper. We rear our lamb on relatively poor land, which cannot compete with sheep reared on the lush pastures of New Zealand. Lamb producers here are very dependent on subsidies – often nearly all their earnings. With the 'subsidise workers not owners' system advocated earlier, they would continue to be subsidised by much the same amount – as they would be working the land.

The state needs to step in, as it did after the Second World War and until the 1980s. Rather than leave it to the market to decide, we should use this opportunity to determine what food we import and what we grow. Clearly there is a real challenge to how much and what sort of foodstuffs we can produce. There are biological and geographic limits. But they are not as limited as you might think. Whenever you travel through the countryside, you can see how differently we can use land. When you see 'ridge and furrowed' fields, these indicate they were growing crops 300 years ago.

Monocultures are efficient but don't produce the variety we could produce. As Tim Benton's report spelt out, we need much greater diversity of foodstuffs and farming techniques. It is harder to harvest different sorts of crops – rather than a two-mile field of celery. Some of the grassland could be cultivated. We should be using rotations much more. 'Mob grazing' is where many cattle feed in small paddocks and are then moved on. This technique reflects a more natural way of grazing and the pasture is better for it.[19] Poorer land could be home to trees and orchards. We could grow more fruit and nut crops on trees way up the hillsides, which would feed us all a lot better than a few grouse. There are many other crops we could produce, which we will pick up on in Chapter 6. We could provide much greater diversity with some imagination.[20]

Clearly other countries grow oranges far more easily than we can. But it is surprising how much we *can* grow. The dominant banana variety in the world is called 'Cavendish' after the family of the Duke of Devonshire, who owns Chatsworth House in Cheshire. The sixth Duke bought some dwarf bananas in 1834 that had come from Southern China via Mauritius, and got his gardener Jo Paxton to grow some in the glasshouse. He produced so many that he gave a visiting missionary some to take back to the Caribbean. The people there must have thought God had smiled upon them, as these dwarf bananas were resistant to a fungal Panama disease that was wiping out the crops of 'Gros Michel' banana that they had been growing. I remembered this story because recently I worked in the Cavendish Building, on the Campus of Manchester Metropolitan University, where I produced a 'Sustainable Food' unit for their new Food Entrepreneurs Degree. History is repeating itself. The same Panama disease is now wiping out the Cavendish bananas. It is being called Bananageddon – because there is little diversity in the crop to develop resistance to the wilt. We need the sort of research the good Duke did, on a larger scale for many more crops.

This tells us clearly that we should not feel limited by what we are doing with the land now. There has been serious underinvestment for many years in land. We should have visions of what we can do that far surpass what we are doing now. Sir Tim Smit has given us the best example of how our land can be transformed. The Eden Project was a clay pit in Cornwall that he transformed into a tropical forest and Mediterranean garden. He said on 'Farming Today' that he was petrified at how we have allowed agriculture to decline as a science, how agricultural colleges find it difficult to attract people as a profession, and that farming as a 'brand' has also declined.

He was especially outspoken on the value of research stations like fruit research at East Malling and famous plots at Rothamsted. I knew both quite well, as Malling funded my masters and I had good friends among

PART II

Society

4
Trade

In this chapter we look at the role of trade in our food supply. Some believe that 'free trade' is the be all and end all, and will solve our food problems. We shall see here that free trade in food is not the answer – but the problem. Time and again we bow down to the god of free trade, which does nothing to help the environment, our health or the long-term well-being of the nation. Increasingly it benefits people willing to gamble on food.

FOOD TRADE DEFICIT

Many people seem to think that the world is our oyster, ready for the picking. I suspect George Cole in the TV show 'Minder' may have been nearer the mark when he quipped 'the world is our lobster', i.e. ready to bite.

For the last 25 years, British food policy has been more or less left to the supermarkets. The result is that food imports amount to $66bn – virtually half of what we eat. And half of this comes from the EU. We export almost exactly half that – $33bn – leaving a $33bn Food Trade Deficit (FTD).

Let's turn those percentages into people. Our population has increased during the same period. About 10 million extra people are now being fed by people and resources from elsewhere. Many of the people in the food chain cannot afford to buy what they grow, so go hungry or rely on food banks. We are using other peoples' resources – particularly land, labour and water – to feed ourselves.

That Food Trade Deficit could come down over the next 5 years in two ways – by increasing food exports or reducing food imports.

The government is putting its energy into exporting more foodstuffs, so we'll look first at their predictions (Table 2).[1] In column 4, I provide a guesstimate as to what might actually happen.

Table 2 UK Government Food and Drink International Action Plan, 2016–20

Where	*What*	*How much £m*	*My prediction £m*
Australasia	Beer and cider	293	200. They already have Fosters
Mexico/S. America	Traditional grocery and alcohol	215	150. They will have cheaper local alcoholic drinks
France	Branded grocery and 'food to go'	132	32.We are more likely to lose lamb (est £100m sales) not gain
Germany	Speciality 'free from' and spirits	610	400. Again will take a lot to make up for loss of beef exports there
India	Tea, biscuits and beer/ spirits	349	250 Based on them lowering spirit tariff, but no sign of it
Japan	Tea, biscuits and seafood	185	150. Can't see seafood sales, but hopeful for all those hobnobs with tea
USA and Canada	Lamb/beef and alcohol	579	400. Can't see their food services buying our beef when they can get cheaper hormone beef. Perhaps lamb OK
China	Meats, premium seafood, beer, seed potatoes, barley grain	405	300. Heard promises before – £45m for pig semen didn't transpire. Making it from meat exports of cheap cuts/offal seems hard work. Hopeful about barley grain
UAE and Gulf	Ex-pats eating more 'free-from' and sweets and snacks	154	100. Only so much ex-pats can eat
Total		**2,922=3bn**	**1,982 =2bn**

The government calculates food exports just short of £3bn – over a 5-year period. I predict they will be lucky to reach £2bn over same period. Most of my predictions are based on lower expectations of their increased export expectations, but some are also due to actually reducing present food exports – in particular meat to the EU. I look forward to seeing how my guesstimates pan out.

The other way to reduce the Food Trade Deficit is by reducing our food imports, and seems a lot less work. The idea is to reduce a percentage of food imports over the same 5 years. I use dollar values in Table 3, as these food import figures are based on international figures from the Observatory of Economic Complexity (OEC) per year.

Table 3 sets out a way to produce savings on food imports of over $3bn over 5 years, which compares favourably with the proposals from the UK government to increase food exports by £3bn. I suggest both need work, but reducing food imports is much more achievable. We should set these targets, and then devise the plans in order to achieve them. Each is quite feasible, and sectors could be rewarded for achieving them.

Table 3 Food Imports (in dollars), Possible Targets as Percentages and Savings over 5 years

Animal Products				*Plant*	*Products*		*Savings*
What	*Import costs in $bn*	*Per cent reduction target as %*	*Savings value $bn*	*What*	*Import costs*	*Reduction target as %*	*Savings value $bn*
				Grapes	1.0	10	0.10
				Tomatoes	0.5	33	0.16
Cheese	2.5	20	0.5	Apples and pears	1.0	33	0.33
Poultry	1.5	15	0.22	Other fruit	0.5	33	0.16
Beef	1.0	20	0.20	Wheat	0.5	33	0.16
Pig	1.0	20	0.20	Corn	0.5	20	0.10
Preserved meat	1.0	25	0.25	Frozen veg	0.5	33	0.16
Fish	1.0	25	0.25	Rice	0.5	20	0.10
Animal Feed/soya meal	0.8	50	0.4	Wine	4.0	10	0.40
Total			**2.0bn**				**1.7bn**

Source: Observatory of Economic Complexity (OEC)peryearhttp://atlas.media.mit.edu/en/profile/country/gbr/#Imports

I show we could reduce imports of meat by 20–25 per cent over the next 5 years. Instead of trying to export meat to the EU, we could eat more of it ourselves, and reduce imports from New Zealand and Argentina. One retailer, the Co-op,[2] has taken the lead, saying that all their meat will be UK-sourced in the future. They believe that this commitment will attract customers – in other words, there is a market. I welcome this, and would welcome it still more if the animals were fed only on UK feed. We import $0.8bn worth of soya meal for animal feed per year. It is probably harder to increase poultry production as much, hence the lower 15 per cent target.

We could grow much more fruit and vegetables ourselves, hence the higher targets – to reduce imports by a third. This would be achievable, provided growers were given guarantees that retailers would buy their products. It would take some consumer education to change from rice to barley and cauliflower. Wine imports would be difficult to reduce, hence the low 10 per cent target. By costing those percentage reductions against import costs, we could 'save' $1.75bn on animal-based products and a further $1.5bn on plant-based products, making a saving of over $3bn on food imports. Whether you agree with these guesstimates or not, the question of reducing imports should be seriously discussed, as an alternative way forward.

By reducing food imports, we not only save money, we create new internal markets. We pay ourselves, and help our own economy. Where else can we create an opportunity like that? Instead of running round the world trying to find international markets, we could create our own internal market – worth more money – here.

We need to have a look at how 'free' food trade may occur in the world in the future. Our free market friends aren't just looking for export possibilities, they are also looking to import more food – more cheap food.

MAD MARKETS

The free traders who want out of both the Single Market and the Customs Union, believe that free trade is the answer to all of our food problems. Yet the history of food – as we saw in the opening chapter – shows that free trade is not the solution; it is a major part of the problem in our food chain.

The UK Tenant Farmers Association Chief Executive said in a speech in March 2016:

> I stand here representing the tenanted sector of agriculture and farm families up and down this country who work hard day in and day out to produce quality food to high standards of animal welfare, environmental management and food safety who feel disenfranchised, marginalised and stolen from. At the same time I am sorry to say that our policy makers are chasing the fantasy that the free market is the answer to our plight.[3]

Free trade does not deliver for food. Relying on free trade provides no food security for the future. Farmers are not only at the behest of the weather, they also have to survive the volatility of the market. Whenever somebody has a good harvest, they think their earnings will go up, but they rarely do. More often they go down, as all the neighbours have a good crop too, so the markets are saturated and prices plummet.

Consider a good apple harvest – the apples get left on the ground. In Kent, in autumn 2014, there was a fabulous harvest of apples, as the winter had been cold enough to kill off the bugs, and the late summer wet enough to fill out the fruit. The orchard owners had a smile on their faces for a good job done. But prices plummeted, as there was a glut on the market throughout the EU.[4] Grain harvests that year were also good – so prices were on the floor.[5] In 2016, the *Wall Street Journal* headline spelt it out: 'US Farm Income to Fall to Lowest Level in Nine Years. USDA predicts a 36 per cent decline amid a slump in commodity prices and another large harvest.'[6]

Many commentators usually ignore this ongoing problem of over-production and saturated markets. Yet food capital has to deal with saturated markets every day, usually by cutting the costs of production as much as possible. Many sectors of industry face saturated markets eventually, then find new goods to sell us. However the food industry faces saturation all the time, as we can only eat so much, and so cuts costs of production even more – hence the high incidence of cheap labour in the food sector – see Chapter 6 for more.

Another way to deal with this saturated market is to stuff more food down our throats. We are one of the most obese countries in the world (although there is now a global epidemic). There are many theories for the rise of obesity across the globe in last the 25 years, but the most convincing is that we are eating too much rubbish. Food capital wants us to eat as much as possible and then bite a bit more.

Another way to sell us more food than we need is to make sure that a lot is wasted on route. An estimated one-third of all food in Britain is wasted. Over a third is wasted in the fields, when supermarkets refuse to buy less than perfect produce, apart from a few select schemes like Morrisons' 'wonky veg' scheme. Another third is lost at home when we've been convinced by '3 for 2' offers that we should buy more. 'Sell by' and 'use by' dates all help to create confusion where once our nose ruled and taste could tell if our food was still good to eat.

Free markets are not the solution, but a major part of problem. We are told 'what the markets want', as if those markets had a mind of their own. Adam Smith said an invisible hand sorts them out and is 'The unobservable market force that helps the demand and supply of goods in a free market to reach equilibrium.'[7] That sounds as real as a magic wand to me.

Mind you, I wish the free traders would follow what Adam Smith actually said. While he believed that rational self-interest and competition can lead to economic prosperity, he also argued that a country's wealth lay in its labour – not gold and silver. Free traders embrace his teachings about removing tariffs and having a smaller state, but forget his enlightened self-interest principle – that it's 'in the butcher's interest to sell good meat at a price that customers are willing to pay, so that both parties benefit in every transaction'. Good meat, not cheap meat.

This is the crazy world of food and farming I captured in *More Than We Can Chew* 30 odd years ago.[8] Quite simply, there is a limit to how

Photo 6 *More Than We Can Chew* was published in 1982.

much people will eat. We were taught in agricultural economics (at Newcastle University) that there are staple foods where demand changes little (demand is inelastic), but other foods where a small price change may alter behaviours, e.g. high beef prices leading people to eat more pork. This is elastic demand. It is elastic that food capital wants – elastic to expand round your widening waist.

Overproduction is the farming norm and the bane of money-makers. Yet we are still encouraged to grow more – more efficiently – and then increase productivity. The price for overproduction is less profit. So farmers have to be more efficient to make up the difference. Only to find everybody is advised to do the same, so profits slump further. So they have to introduce even greater efficiency. The only way out is to do something else – diversify. If we grew lots of different crops and ate lots of different foods, we could avoid many of these saturated markets. If we stopped importing so much food, we would create larger internal markets and better prices for food producers.

GLOBALISATION

The recent Brexit and Trump votes are being put down to a feeling of 'being left behind' following globalisation. Sir Simon Fraser said to the Worshipful Company of World Traders (I kid you not) in the Tacitus lecture entitled, 'The World is Our Oyster?', in February 2017 in the City of London:

> The disruption has come not from the world's most disadvantaged, nor from countries people here label as protectionist, but within the two rich countries that have been the most vocal champions of open trade in a rules-based multilateral system: the United States and the United Kingdom. We now have an American President who says 'protection will lead to great prosperity and strength', while Britain is leaving the European Union … The EU single market is by far the world's most successful example of international trade liberalisation and regulatory cooperation, and was our brainchild.[9]

I couldn't want for a better source than this lecture.[10] Many excluded from the process of globalisation found a voice in the form of Brexit and Trump's election, and the June 2017 election. People have felt 'left behind', and have yet to find a new identity. Building shopping centres goes only so far. People have been squeezed out of secure work by outsourcing or cheaper foreign labour. Many are stuck on zero hour contracts for dead-end jobs. This is all especially prevalent in the food and farm sectors, as we'll see in Chapter 5. The process of globalisation has accelerated, driven by technological change, which feeds frustration, grievance and hostility to immigrants. It is hard to see how it can be replaced – other than growing our own food and economies, and revitalising our cultures through food.

WORLD TRADE ORGANISATION

We hear people saying 'It won't be the end of the world coming out of the EU. We can trade freely with who we like, according to WTO rules.' Nothing could be further from the truth. Everywhere there are taxes on the movements of foodstuffs. We saw early on how complex the EU tariffs are on 2,000 agricultural products, many with different rates for different formulations, varying from country to country, meaning all 164 WTO member states. These are a result of years of complex deliberations, to protect various sectors in each country and designed to be popular with their own inhabitants. It is hardly a nice smooth pond we are sailing into – more like a shark-infested quagmire.

It is not clear what the status of the UK would be after coming out of the EU without a trade deal. According to the *Economist*:

> Britain would need to have its own 'schedules', WTO-speak for the list of tariffs and quotas that it would apply to other countries. In theory, it would not be too difficult for Britain to acquire its own schedules. Under a so-called 'rectification', the British government would simply cross out 'EU' at the top of the page and write 'United Kingdom' instead. Doing more than this would be difficult. Imagine that Britain decided that it wanted to boost support for its farmers by raising tariffs on farm products (perhaps they had lost out from reduced EU funding). Such a move would require a more ambitious 'modification' of Britain's WTO schedules, requiring much lengthier negotiations. While they were going on, Britain's status in the WTO would be in legal limbo.[11]

Tariffs are not confined to the EU. They are everywhere. The starting point is that food is rarely traded 'freely'; it is usually 'taxed'. Every country has reasons to protect some of its food and farm producers. If they don't, they don't last long. The easiest way to put importers off and protect your own suppliers is quite legitimate and it is to introduce a tax on their foodstuffs.

The WTO says:

> The rate of duty depends on the Harmonised System (HS) of product classification, a system developed and maintained by the World Customs Organisation (WCO), in which traded products are assigned internationally standardised HS codes and commodity descriptions… The rates of duty may differ depending on the country of manufacture of the imported goods. Preferential duty rates may apply where a Free Trade Agreement exists between two or more transacting countries.[12]

The WTO came into being in the 1995 and replaced GATT – the General Agreement on Tariffs and Trade – after years of multilateral negotiations. It is not part of the UN but a members' club in which members agree trade rules, and give the WTO powers to adjudicate in disputes and impose sanctions. While agricultural products and foodstuffs are not the largest trade category, they have had more than their fair share of disputes. There have been about 100 agricultural trade issues submitted to the Dispute Settlement Procedure in the first 20 years of life of the WTO.[13] Many are to do with health and safety, despite clearer rules on sanitary measures. There are also increasing challenges about subsidies, as countries prefer to litigate rather than negotiate over them.

There have been several long-running disputes between the USA and the EU, particularly over bananas and hormones in beef. The US wanted to strengthen the disputes process in order to settle these.

There was a spate of challenges to do with quotas in the late 1990s. Brazil challenged the EU on quotas over poultry and New Zealand challenged the EU on butter quotas. The US challenged the Philippines on pork and poultry quotas. In 2002, Australia, Brazil and (later) Thailand challenged the EU sugar regime. The conflict was over the extent to which the EU provided export subsidies (to companies like Tate & Lyle). In 2005 the European Commission agreed to cut subsidies to reduce sugar production and thus the need for export payments. In 2003 the US challenged the EU's system of protecting Geographical Indications (GIs), which had implications for a century-old dispute regarding the brand names Bud and Budweiser.

About half the disputes are bought by just four WTO members – the US, the EU, Brazil and Argentina. It had been expected that it would be developing countries bringing disputes against developed countries, but it is largely the other way round. 23 smaller countries have asked for consultation – the first step in the disputes process – but have not gone further.

The WTO rules always favour freer market access. They don't like barriers to trade, whether monetary or sanitary. Except those like quality standards.

One of the more bizarre disputes was when the US challenged India. India has a $12bn food security programme, a key welfare measure aimed at delivering millions of people from poverty. India buys grains such as rice and wheat from farmers at above market prices, and stockpiles to protect against shortages. The WTO Agreement on Agricultural Subsidies allows only 10 per cent (of production) subsidy for most developing countries. The US argued that India had exceeded this 10 per cent and therefore 'distorted' international trade. Something called a 'peace clause' was in place to allow food security measures like this, but the WTO decided this would only last four years. India dug its heels in, saying it wanted the peace clause to last indefinitely, and if it didn't, it wouldn't enter a major new trade arrangement expected to generate $1 trillion. India got its way in 2014.[14] But it makes you wonder what the world is coming to when a country can't protect its own people like this.

It is a curious that in the wonderful world of 'free trade', there are tax barriers everywhere. Free Trade Agreements are about avoiding these taxes. We can go anywhere in the world now and trade 'freely' – we just have to pay the taxes. Because we are so dependent on overseas food, we will be more at the mercy of international markets once we leave the EU, so we will need trade deals to include foodstuffs. The dominant players in all food markets are USA, Russia, China and the EU. Our food trading outlook has to start with Europe, then the USA, then the rest of the world.

EU SINGLE MARKET

In 2015, 44 per cent of total UK food and farm exports went to the EU, and 53 per cent of UK imports came from the EU. None of the possible forms of Brexit can give us the same access and influence in the EU market. We heard from the Brexit Minister, David Davis, on the first day of UK-EU Brexit negotiations, that we will get a 'deal like no other'. It is hard to see how any deal can cope with 2,000 agricultural products and 15,000 PAPs. One trade deal involving a few commodities can take years to agree, so a deal with these thousands could take a lifetime. Without a deal it is hard to see how we are going to move as much food around the EU. And priorities for the deal 'like no other' lie elsewhere.

The government has mentioned in a White Paper the possibilities of sectoral agreements with the EU. The danger for food and farming is that they may be marginalised as the financial and automotive industries would have priority in any EU-UK trade deal. This was evidenced by the Chancellor of the Exchequer, Philip Hammond, in his speech to the Mansion House on the day negotiations started. He didn't mention food and farm services, but said: 'Britain must secure a good trade deal for financial services post-Brexit.'[15] He went on to chide EU counterparts for 'protectionist agendas being advanced, disguised as arguments about regulatory competence, financial stability, and supervisory oversight'.

The chance of a deal like no other for food and farm goods looks slim. The UK relies far more on food exports to the EU's internal market than any individual EU country relies on food exports to Britain as part of the internal market. Such a weak bargaining position is unlikely to play well, despite David Davis's negotiating tactics.

Sir Simon Fraser at that Tacitus Lecture said:

> There is a serious risk during the Brexit negotiations. If the Article 50 exit negotiation does not go smoothly, our future trade relationship will be negotiated not from the starting point of the status quo – integrated membership of a common market and regulatory space – but from outside, almost like any other third country. Negotiations will cover many issues: tariffs and other barriers for trade in goods, market access and regulatory equivalence for provision of services, mutual recognition of product standards, dispute settlement and more. Take the customs union. Leaving it would imply additional tariff costs on goods. EU tariff for food and farm stuffs are often over 10%. We should aim for an FTA agreement to reduce most tariffs to zero or near zero, just like the EU has recently agreed an FTA with Canada. However, under WTO rules we cannot unilaterally offer a zero-tariff to the EU unless we also offer it to all other countries – the trade equivalent of unilateral disarmament.[16]

He went on to say that tariffs are not the critical issue, because the added costs of non-tariff barriers could be between 5 to 10 per cent. It is the cost of customs checks, inspections, regulatory compliance, more intrusive paperwork or delays in distribution that will obstruct free movement of foodstuffs. This will increase, not reduce, red tape and we will have to replicate 34 EU regulatory agencies to comply with the rules of movement.

DOING DEALS ELSEWHERE

Wherever we go to try and do new deals, we will face existing tariffs. That is the whole point of a deal – to get rid of them. Those tariffs are there for all sorts of reasons that the host country values. If we are to do a 'deal' we will have to offer something in return. Each deal will be a two way process. The trouble is that – in general – we will want them to remove tariffs on our financial services, white goods or cars. And occasionally, whisky. The main trading contenders, the USA, Brazil, Argentina, Uruguay, Australia or New Zealand – will want our tariffs on food stuffs reduced or removed. While each of those deals may be good for our finance or manufacturing sectors, they will be bad for our rural economy.

Don't get the idea these deals will be made overnight. The multilateral deal called the Doha round in the WTO is stuck after 16 years. The bilateral deal between EU–Canada took 7 years to negotiate. The US prides itself on much faster bilateral deals – taking 18 months on average,[17] hence President Trump's preference for bilateral deals. A New Zealander, Crawford Falconer, has been bought into the Department of International Trade, based on his experience negotiating deals for New Zealand with Asian neighbours. The one with Hong Kong took 10 years to come into force; however, Malaysia took only 5 years to negotiate while China will take 15 years for full implementation.

The first priority post Brexit should be new arrangements with the group of 50 countries with which the EU has FTAs. They account for 13 per cent of current UK trade, of which 70 per cent is with just 4 countries: Norway, Switzerland, Korea and Turkey. There are also countries with which the EU is currently negotiating, that would increase the total to 25 per cent of current UK trade. When we leave the EU, we will cease to have privileged trade access to these markets unless by mutual agreement we continue the terms of the EU FTAs or negotiate new deals. Otherwise trade would cost more and the UK would be at a competitive disadvantage. The Korean FTA, like the Canada-EU FTA, is doing away with most of the tariffs.

United States of America

The second priority has to be the USA. In July 2017 Donald Trump said a US-UK deal could happen quickly. In 2015, 20 per cent of total UK food and farm exports went to the US, and 11 per cent of our food imports came from the US, amounting to around 16 per cent of our total trade. The US has long been an ally, we speak almost the same language, and a small increase in percentage trade is massive in actual amounts. But we can't start deal-making with them now, until we have sorted our deal with the EU.

The US will be eager to get rid of their ever-present food surpluses. These will come in the form of chlorinated chicken, hormone beef and GM corn (see Chapter 11). But trying to guess what President Trump may do is proving tricky. We would have expected him to repay all those farmers who voted for him, but so far he has done that by reducing subsidies to the agricultural insurance scheme.

The issue of plant ownership would certainly be part of any US deal. The EU is satisfied with Plant Breeders Rights (PBRs). PBRs enable people to buy plants and keep whatever they produce. They cannot sell them on as their variety – the plant breeder owns that right – but they can use them to develop new varieties. The US prefers 'Patents', which come from the chemical industry and are much more restrictive, as we will see in Chapter 10.

BRICS

Then there are the BRIC countries. Many commentators say that this is where we create new exports to compensate for the ones we've lost in the EU. Dynamic markets like the BRICs – Brazil, Russia, India and China – are seen as good prospects. In 2015, 8 per cent of UK food exports went to all the BRICS and South Africa combined, and 11 per cent of our food imports came from them.

Brazil

Brazil wants to feed the world market and has capacity to spare. However, much of the land being taken over to grow soya is cattle pasture. While some might think getting rid of cattle to grow soya is a good thing, it is not good for the soil.[18] Most of that soya goes to feed the Empire of the Pigs (namely China).[19] Some of that soya will feed Chinese pigs, but most will be deployed to keep food prices low.

Several ports are opening up in Brazil to ship soy. The great grain merchant, Bunge, opened a $700m soya port in Barcarena, at the mouth of the Amazon River, in 2014. Bunge is one of the big four *Merchants of Grain,*[20] now called the ABCD companies: ADM (Archer Daniels Midland), who are building a soya protein concentrate plant in China; Bunge; Cargill and Louis Dreyfus (both essentially family-run businesses).[21]

The shift of Brazil's economy towards agricultural commodity exports is one of the great food trading movements in the world now. It strengthens the influence of large landowners and moves away from manufacturing – the opposite of everywhere else – weakening the influence of industrialists and labour unions and affecting virtually every aspect of Brazil's politics. I await the slump in soya bean prices.

Russia

Russia is producing more grain for export than ever before, and is likely to continue, now it has taken over the Crimea, probably the most fertile land in the world. From supplying a few per cent of the world's grain in 2000, it now accounts for around 15 per cent. Putin will have plans to use this grain as a negotiating tactic – seeing food as power, as the US 'Food for Peace' programme did over many years.

India

Despite talk about prospects with India, there isn't a lot of food trade with India and few prospects to expand. The highest tariffs faced by the UK's exports into India are on beverages and spirits, followed by coffee. There are high tariffs on dairy products,[22] which have prevented a deal between India and the EU, so it is difficult to see why we would be able to get a better deal. UK exports to India as a percentage of India's total import market fell by half in the 10 years to 2015, losing market share to Germany, France and Italy.

China

I'm not so sure about China when we go round the world looking for food deals. They have much greater need and power. Their greatest food need is to feed their new workers cheaply. There is internal migration within China of around 200 million people from the countryside to the cities to work in the factories. From feeding themselves on a healthy diet in the countryside, most of these young people are now picking up many of our eating habits. I went into a Tesco store in China and the shelves were laden with sweets and chocolate, the like of which we wouldn't allow here.

Travelling through China I saw the countryside being run down and overgrown with bindweed. Driving back into Guangzhou (Canton), with a population of 13 million, I saw hundreds of young people in work uniforms, pouring out of an electronics factory. These new workers have to be fed very cheaply. Otherwise the products we buy from them won't be cheap any more.

China has built the largest port (£1.6bn) in the world in Brazil, called Superporto do Acu, to move millions of tonnes of iron ore, grain, soy and millions of barrels of oil. (It is ironic that the Chinese are importing so much soya, as soya was first cultivated there, showing again how most crops are grown on the other side of the world from where they were first cultivated.[23]) Soya does not attract any EU tariffs, but I can't see us competing with Brazil to export soya to China.

We may be able to sell more of our meat offal there – but it seems a long way to go to get rid of our meat waste. The UK government hopes to increase barley sales to capture a newly-emerging beer market there, and there could be some mileage in that. However, exports to China have barely moved in 10 years, despite a recent deal to buy barley.

If we were to lose 5 per cent of our trade with the EU, in order to make up that volume and value, we would need a 25 per cent increase in our food trade with the BRICS and the Commonwealth.

Commonwealth

In terms of foodstuffs in 2015, Commonwealth countries accounted for a mere 9.5 per cent of our food exports and 8 per cent of our food imports. Australia accounts for 1.6 per cent of our exports and provides 0.8 per cent of our food imports. New Zealand accounted for 0.2 per cent of both exports and imports. While desirable, it would seem to be taking up much-needed negotiating resources we can ill afford because although FTAs with New Zealand and Australia would look good, they would account for little in the way of food exports. A further 30 Commonwealth countries, mainly in Africa and the Caribbean, are already covered by EU FTAs or have tariff-free access to our markets. This raises two issues. First, until we negotiate new agreements with them, we risk being in the odd position of having worse trading terms with these Commonwealth countries than the EU. Second, the UK will be unable to champion their access to the EU market, as they already have that.

We could develop new food trading relationships with Commonwealth countries in Africa, but the uncomfortable legacy of British colonial rule in Africa persists, and could be problematic. There are memories of colonial legacy all round, which I ran into many times while working in Commonwealth countries. These include what has happened to Tamil people, taken by the British to work the tea plantation in Sri Lanka, and the role of the Kikuku people in Kenya, left the best land by the British and still in power today – despite a blood bath following elections in 2007. Whether we can walk in and do deals, as if nothing happened, is doubtful.

While we are hyping up more trade from Kenya and the rest of Sub-Saharan Africa, we should also factor in a few environmental constraints. The Sahara desert is growing by about a mile a year. That desert was not desert 2,500 years ago – the land was wrecked by the Romans. So the desert is not 'natural', and I would argue – as did Ritchie Calder before me – that if we could do any one thing to bring peace and reduce gross inequality it would be to replant the Sahara.[24]

However, we are making that task more difficult by the year. As I write this, I'm bombarded by appeals to contribute to help the dreadful drought across the whole of East Africa, including Kenya. We are told it is the worst famine crisis since the UN began 70 years ago, exacerbated by 'conflicts' and 'corruption'. It is stirring up old divisions, as pastoral farmers, who move their cattle around, invade conservation areas which have better vegetation, because they are well managed with overseas funding.

None of this is being helped by exporting water from places like Kenya, in the form of food exports. Kenya's two major food export crops are tea and coffee. These crops need a lot of water to grow (see Chapter 7). It takes more than 1,100 drops of water to produce one drop of coffee. Coffee in dollar terms is the most traded agricultural commodity in the world. To drink coffee, the world's population requires the equivalent to 1.5 times the water streaming out the River Rhine every year. Of all the water needed to produce crops and creatures sold in the world, about a tenth goes into international trading.[25]

Many African producers have already set up cooperatives and trading organisations, making direct deals harder. The Tanzania Pigeon Peas Cash Crop, for example, sell to the Exporting Trading Group, which ships millions of tonnes of commodities. They have set up sales arrangements already – they do not need any 'deals'. A one-acre plot now links to the world – through numerous massive trading groups. However, these small farmers are still utterly dependent on selling their raw commodities. Rather than looking for deals, we should help them build up the food capital near their farmland, where local food-processing plants could add real value.

FOOD FUTURES MARKETS

With all those tariffs and few prospects of any Free Trade Agreements in the near future, we may be a bit more dependent on buying food in the global market place. There we will be at the mercy of volatile price fluctuations. In which case, we will need to venture into the food futures markets. In the mid-twentieth century something called future trading was set up to try to stabilise matters for those in the food chain by agreeing prices in the future.

The futures market allows producers (or anyone in the farm/food chain) to hedge their bets by making a contract with a dealer to ensure a decent price for their food produce. Around 20 years ago, the futures market was opened up to an increased number of foodstuffs, and allowed anybody (not just those in the food chain) to speculate on futures contracts.[26] The Clinton presidency encouraged this 'liberalisation'. This encouraged many countries to buy from global markets, rather than relying on regular sources, as they could get cheaper supplies that way. Arab countries imported 40 per cent of their food by the 2000s. We are even more dependent on food imports.

Where coffee manufacturers could once buy 'future' coffee to cushion themselves from sharp price rises, now speculators can make money whenever they see a chance. These futures food markets have turned into a casino. Today, around 97 percent of food futures trading is done by speculators in foodstuffs. They bet on livestock, like pork bellies, and on orange juice and coffee. You don't need any food or farm skills to make a killing.

Futures food markets enable a lot of gambling on foodstuffs, by people not in the food chain. Gamblers look for where there may be shortages, due to adverse weather like frost or floods in order to bet. But food is a major commodity in markets round the world. We could find ourselves in a vulnerable position – at the mercy of speculators.

The head of the Food and Agricultural Organisation (FAO) said in 2012: 'The world needs to take a hard look at speculation on the financial markets and its potential impact on food price volatility. While there has been much analysis of food price volatility, more understanding is still needed, especially concerning the impacts of speculation. Excessive food price volatility, especially at the speed at which they have been occurring since 2007, has negative impacts on poor consumers and poor producers alike all over the world.'[27]

Only about 2 per cent of goods traded on the futures markets actually ends up with the purchasers, leaving more room for speculators. According to Global Justice, financial services have spent billions of euros trying to persuade governments not to restrict such speculation.[28] The US introduced laws to curtail it,[29] and the EU wants to as well. Nicholas Sarkozy, then Head of the G20, called for tighter controls ahead of a meeting of G20 agricultural ministers looking at greater investment in agriculture. But the then UK Agricultural Minister, Jim Paice, said there was little value in tighter regulation and the then PM Cameron complained that it would stop the City making money.[30]

We cannot leave it to people making a quick buck out of food. It may provide opportunities for financiers to cash in, but does little for workers and consumers. Banks are earning huge profits from betting on staple food prices in unregulated financial markets. This creates instability and pushes up global food prices, leaving millions hungry and facing deeper poverty. In January 2014, the EU agreed to introduce new rules to prevent hedge funds and investment banks from driving up food prices.[31]

In 2016, the NFU had to lobby to enable farmers to stay in the food futures markets. They called on MPs to avoid the unintended consequences of a new EU directive to stop gambling in food markets: the Markets in Financial Instruments Directive II (MiFID II). This directive threatened farmers' ability to use the futures market for selling their own grain, by regulating farmers as if they were in the financial sector.[32] It would indeed be ironic if the finance houses kept the farmers out of food futures markets that were originally designed to help them.

Food Prices

Food prices in the UK have gone up steadily since Brexit. It started when Unilever tried to hike the price of Marmite, but price rises in all food have been inexorable, mainly because of the drop in the value of the pound. In the first three months of 2017, food prices rose more than in the previous three years, despite retail wars keeping prices as low as possible.

City[33] analysts predict that food prices will rise by 8 per cent due to Brexit itself. Fruit and vegetables, flowers and olive oil will all become more expensive once the UK leaves the EU, no matter what trade deal is struck. This will be the cost of 'control over our border' as Britain is forced to impose extra border controls.

The *Dispatches* Programme on Channel Four, in February 2017, showed that cheese producers face a hike of 30 per cent on costs,[34] because the powdered milk they use to make cheese, is traded in dollars, so their buying power has decreased dramatically.[35] Harry Smit, a senior analyst at Rabobank and author of 'Future Food Security in the UK'[36] said in April 2017 that the UK needed to brace itself to pay more for fruit and vegetables from Spain and the Netherlands and wine from France. With such barriers, however, it does mean that there are incentives to produce more raw food stuff ourselves, to supply our own food manufacturers.

E-FOOD TRADING

We shouldn't forget e-commerce in food markets. Already African farmers use mobile phones to find out current market prices for their food. However, they are still at the mercy of these prices. The Department of International Trade has set up an e-trading programme to help UK companies sell their goods and services to millions of global consumers through online exports.[37] However, I think we should go much further, and be using our e-technologies to help them develop food manufacturing and services. See Chapter 5 for more.

Seven Cultures of Capitalism, by Charles Hampden-Turner and Fons Trompenaars,[38] examines different business cultures in Japan, Germany, France, Sweden, the Netherlands, the UK and the USA. It reveals the strengths and weaknesses of the individualistic and communitarian approaches to capitalism, and provides us with options as to the sort of Britain we may want. It provides insight into wealth creation and business thinking around the world. One of the insights is how Japan creates wealth in quite different ways from Britain and the USA. There they talk about products increasingly forming 'food chains'. It is not a matter of inventing a product that has a life-cycle (beginning, middle and end), but of continuous incremental developments. What used to be a race is now a relay race, where you depend on the previous runner to pass on the baton. The faster your team-mate hands over, the better. Everywhere teams are replacing individuals.

We have seen some of this in the food chain itself in the past few years. Supply chains for both coffee and beer have been transformed. Coffee shops have become transnational enterprises, forming part of a whole new culture. Craft beer producers, using newer hop varieties, have established microbreweries and new styles of outlets, in the process passing the baton on to the next link in the chain. The export of our service industries nearly doubled in the 20 years from 1995. Imagine what our food services expertise could offer other countries. We could export our food service sector across the world, enabling food entrepreneurs to develop in Asia and Africa, and making these countries less dependent on corporates in the process.

I distinguish corporates from companies, to indicate that not all capitalism is the same. We have seen how corporates have more power than many nation states. A friend who worked for one – as their ethical manager – told me he would just ring a government official in – say – Kenya, and tell them what his company wanted. Corporates can move money around transnationally, thus dodging taxes. I want to develop local companies – whose name means 'breaks bread together' – to deliver this new food sector in developing countries.

5
Labour

FRUITS OF OUR LABOUR

In this chapter we examine what is usually overlooked – labour in the food system. Our food does not reach us by magic. It is delivered by people. For many consumers, this goes on invisibly, but not for those actually on the land and in the food factories, many of whom are badly paid and working in poor conditions. If we want to revitalise our food system, we need to recognise and reward labour better.

Possibly the most meaningful job on earth is planting and ploughing and producing a rich harvest. In the Valley of the Workers in Egypt, the workers painted their own tombs and the images suggest that their idea of heaven was working the soil. A tragic irony is that working the land in the twenty-first century is back-breaking, soul-destroying and badly paid.

Farm and food workers have always been poorly paid, from the Tolpuddle Martyrs, through abattoir workers in Upton Sinclair's *The Jungle* and the heroes in Steinbeck's *Grapes of Wrath*, to present-day deliveroo cyclists. It is the same the world over, according to the many campaigns fought by the IUF – the International Association for Food and Allied workers.[1]

Casualisation – where there are no guarantees about the future, pensions, sick pay, etc. – is becoming the norm in many work areas, particularly the food sector. We hear about zero hour contracts in the 'gig' economy, where workers dash around doing jobs, but are considered 'self-employed'. They have no employment rights, which we have fought years to achieve. Most employment rights – like the Working Time Directive limiting the number of hours worked – derive from the EU, so there will be even less protection on Brexit.

Wigan, a food manufacturing centre in the North-west with over 50 food companies, is a microcosm of the range of working conditions in the food sector. Brewery workers will be well paid, and the cut meat factory offers living wages. But the ready-made meal factory, supplying the likes of Tesco and Iceland, 'interviews' agency workers by dressing them up in protective clothing, putting them on the production line for half an hour, and if they aren't fast enough, sending them home. And at a famous beans factory, one person worked for 9 years on a zero hours contract, never missing a day, only to be told – with no notice – he wasn't required anymore.

MIGRANT WORKERS

'Food *could* rot in the fields' was the cry from the National Farmers Union (NFU) in December 2016. They warned that unless action was taken to deal with the migrant labour shortage, crops in Lincolnshire fields could rot in the ground. Lincolnshire produces nearly 30 per cent of England's vegetables and salads. The prospect of stricter immigration controls has left farmers concerned over an ever-expanding labour gap, despite Lincolnshire being a fantastic county for producing food.

A couple of months later the slogan had become 'Food *will* rot in fields'. Meurig Raymond, President of the NFU, said: 'The U.K. agriculture industry will come to a standstill if the government doesn't reach a deal that guarantees access to European workers. Without a workforce – permanent and seasonal – it wouldn't matter what a new trade deal looks like. The lights would go out in our biggest manufacturing sector, food will rot in the fields and Britain will lose the ability to produce and process its own food.' One in five growers said that they did not have enough migrant workers for the 2017 season and that recruitment was at its worst since 2004.[2]

We did not hear about hard-working migrant workers like this before the Referendum. A letter was sent to the government late in 2016 trying to persuade it not to go for a 'hard Brexit' – not to come out of the Single Market altogether. 'Migrant workers and tariff-free access to the single market are vital for the industry', Morrisons, Sainsbury's, Marks and Spencer, the NFU and other signatories warned. Access to seasonal and permanent employees from overseas was 'essential' to the food supply chain in the UK, they said. In the next breath: 'For our sector maintaining tariff-free access to the EU single market is a vital priority. It is where 75 per cent of our food exports go, so all our farming and food businesses wish to achieve this outcome.' The letter was signed by the NFU in England, Scotland and Wales as well as the Ulster Farmers

Union and 71 food businesses with a collective turnover of over £92bn, including Dairy Crest, Arla Foods, Weetabix and Muller.[3]

I heard not a tweet from any of these characters about the abolition of the Agricultural Wages Board (AWB), which we will find out about later in this chapter. The same NFU and the Fresh Produce Consortium wanted to pay migrant workers only the minimum wage, rather than the slightly higher AWB rate. Apparently it was essential to their business to pay as little as possible – to provide retailers with cheap food. Their mistake is coming home to roost, as migrant workers are now leaving our fields.

British Summer Fruits announced that strawberry prices would soar if they were denied access to their migrant workforce. This threatened the supply of a fruit whose sales had increased by over 130 per cent in the last 20 years to be worth £1.2bn, 'helping our balance of payments'.[4] The organisation warned that the price of strawberries and raspberries could increase by a half and, as they account for nearly a quarter of all fruit sales, this could hit the 5-a-day healthy eating campaign. The chair of the organisation said he couldn't believe that people who voted Brexit really wanted to see the demise of such an iconic horticultural industry and the end of buying British.

A YouGov poll,[5] released the day the letter triggering Article 50 was delivered, revealed that people think:

> It is important that Britain has full control over immigration once we leave the European Union, something that is normally seen as being incompatible with single market membership. The public would like to have their cake and to eat it. When asked a direct question on which they would choose – 16 per cent said control of immigration, 24 per cent said free trade with Europe, but 40 per cent said it was a false choice and Britain could have both. It looks like they will be disappointed.

According to a report from Sustain/Food Research Collaboration (FRC), in 1980 about 5 per cent of farm workforces were 'casual'; that had risen to 15 per cent by 2015. This is part of a general trend towards more casualisation, where employers are reluctant to take on permanent workers.

The food sector employs 3.4 million people (3.9 million if we include agriculture, fishing and self-employed farmers). That is about 12 per cent of all UK employees. The various sectors have changed over the years.

In 1960, farming made up 3 per cent, food manufacturing 35 per cent and food services 50 per cent of the food and farm sector. By 1980, that was 2 per cent, 20 per cent and 60 per cent food respectively. By 2000 farming was down to 1 per cent, manufacturing 15 per cent and food services up to 75 per cent.

But it is not just casual workers who are EU migrant workers, nor just those in the fields. Over the years, many migrant workers have become skilled and taken on positions of responsibility as supervisors and farm managers, all from Europe. Go into any abattoir, and most workers will be from the EU or Africa, or, in the case of Halal abattoirs, of Asian origin – but virtually no 'British' workers.

Unions find these field and food workers hard to organise, although early in 2017 McDonalds agreed to offer all of their 115,000 employers the chance to have regular hours instead of the notorious 'zero contracts',[6] after a campaign by the GMB (General, Municipal, Boilermakers and Allied) and bakery workers' (Bakers, Food and Allied Workers) unions in the company. The company argued that its employees preferred the flexibility of zero hours, but it seems that at least 20 per cent preferred some sort of fixed hours – from 4 to 40 hr/week.

In 2015, the industry with the highest share of foreign-born workers – 41 per cent – was food products manufacturing. 65 per cent of all workers in food manufacturing in Northern Ireland are EU workers. The largest supermarket poultry factories in the UK are increasingly dependent on low-paid EU migrants employed through agencies. The sector is dominated by three companies: Grampian, Moy Park and Bernard Matthews. Moy Park is the largest company in Northern Ireland, and is one of the top ten food companies in the UK, supplying branded and own-label products to a wide range of leading retailers and food-service providers and employing many EU migrant workers. They have around 35 million birds on the ground at any one time in Northern Ireland.[7]

In 2007, Unite accused the poultry industry in general of using agency labour to bypass normal employer obligations in its campaigns 'Make every worker count'[8] and 'Look behind the label'.[9] They gave examples to the Department for Business Enterprise and Regulatory Reform, citing Grampian and Moy Park, where they said over a third of agency workers worked for more than six months but were paid less than full-time workers for the same work, and lost out on sick and holiday pay.

Grampian challenged the allegations, saying its chicken business employs almost 4,000 people, of whom only 6.7 per cent is agency labour.

Migrant agency workers are said 'to complement its core workforce at times of peak production in areas where it has become difficult to recruit'.[10] Some 1,400 union members took strike action in 2007 at Grampian factories over pay, and the use of agency workers. Moy Park also said that it met all the requirements of EU law and ethical trading codes. Presumably they were referring to the Ethical Trading Initiative (ETI) Base Code,[11] run by employers, unions and other concerned organisations. The ETI Base Code is based on the standards for working conditions spelt out by the International Labour Organisation (ILO). It covers matters such as freedom of association, discrimination, child labour, living wages and working hours. I used the ETI base code to produce a web site for 5 major food retailers to communicate with their suppliers.[12]

Unite was one of several unions to force a resolution at the Labour Party Conference in 2007. The Labour government was adamant that the flexibility provided by agency workers was vital to Britain's economic success and took the lead in blocking a new directive in Europe on agency working.[13] Unite said that the plight of more than a million agency workers was 'the single biggest issue facing the country today' and was causing serious racial tension. How prescient this was!

Since then we have seen even more widespread use of zero contracts and flexible working. Deliveroo and Uber come to mind. Without the EU, it is unlikely that we will have any regulation to control agency workers. We have regulations for the way we treat chickens, but not for poultry workers.

Numbers of migrant workers are hard to pin down. David Camp of the Association of Labour Providers estimates that 90–95 per cent of the seasonal workers his members hired in food-processing and agriculture were from other EU countries, mainly Romania and Bulgaria. The most recent statistics have seen a year-on-year increase of 82,000 workers from those two countries, with an increase from 204,000 at the end of 2015 to 286,000 at the end of 2016.[14]

A sign of what may happen was when Romania and Bulgaria became full EU members. Concordia, one of the biggest suppliers of seasonal workers, said that, with free movement, these workers would now go to work in coffee and wine bars, where working conditions were more congenial. There were pressures on the UK government to open up easier migrant routes from Russia, Ukraine and Turkey.[15] They could do the same now.

The Conservative manifesto of 2017 said that in order to bring migrant numbers down they were going to increase the tax employers paid when bringing skilled workers in from non-EU countries, and that money would be used to train British workers. Clearly this will not affect the employment of EU migrant workers in the fields and food factories.

PLANTATIONS

We saw in Chapter 1 that the initial impetus for Brexit came from the Eastern counties, around Boston, Lincoln and Peterborough, who voted 3–1 for Brexit – the highest in the country. When we buy our home-grown food, most of us don't think we are doing so as part of a wider issue. We want cheap food, but we don't want to know why the costs are low. We don't see migrant workers getting up at all hours, working in all weathers, for wages we wouldn't get out of bed for, nor the impact this has on local communities and local labour. Concerns among local people were ignored for many years, making them ready to believe UKIP – rather than do something about the way we produce our food.

There are two sorts of crops grown on arable land. There are the grains like wheat and barley, and oilseed rape grown on large farms, usually with up to half a dozen permanent workers. The other sort is called horticulture. Vegetables like celery, carrots, peas, beans, salad crops and soft fruit like strawberries. It is these monocultures run by migrant workers that I'm calling plantations.

Not enough has been made of the different farm systems in the UK. In the East, arable monocultures dominate – either grain with permanent workers or with vegetable – dependent on migrant workers. The latter system is wrecking not only the land but also local communities, compared to family farms with local permanent workers contributing local food to local communities in the West. The plantation owners were behind getting rid of the AWB (see later), but the consequences have been felt by farmworkers all over the country.

I use the term plantation as it better explains the mode of farm production that is now used to produce much of the soft fruit and vegetable crops in this country. In the past, it was used in relation to tropical crops – not temperate ones. Yet I feel this term better expresses what is going on – rather than 'large farms'. Large farms usually grow grain like wheat and barley and oilseed rape, and may have up to five permanent workers, who can look after over 1,000 acres. Plantations employ migrants to work monocultures of vegetables like cauliflowers, celery and asparagus.

The owners of the plantations – monocultures worked by migrant labour – are pushing for a scheme to allow a similar number of migrant workers as now. It would be like the old Seasonal Agricultural Workers Scheme (SAWS), where limited numbers of Bulgarian and Romanian workers came to work for six months then returned home. The then Home Secretary Theresa May did away with the scheme. There were attempts to trial a new Seasonal Agricultural Permit Scheme (SAPS) in Kent in 2017.[16] While it would look much the same to local communities, the government could say a similar scheme was 'controlled'. It would be more than ironic, if, after all the thousands of laws that have to be changed, there were still many migrant food and farm workers.

When studying for my MSc in Plant Science at Wye College, I argued with my Professor, J.W. Purseglove, about why various crops grew where they did. He was author of *Tropical Crops*, Vols 1–4, and previously the Director of Tropical Crops in Trinidad. He enthusiastically showed how most crops now grown for export, do so on the other side of the world from where they were first cultivated. Look at coffee – first cultivated in East Africa now grown for export primarily in South America.

I found a great book – my botanical bible – called *Plantation Agriculture*[17] that gave me the case for looking more at the role of labour in explaining what crops grow where. A classic is tea in Sri Lanka. I was told tea was grown there because there was a failure to grow coffee because of disease. Yet my botanical bible said it had more to do with the cheapness of labour for coffee-growers in Brazil. There, the plantations took on Portuguese migrant workers who moved around each year. In Sri Lanka, the migrant Tamil population were resident and considered 'more placid' – better for the dull picking work in tea plantations.

Another character around Wye at that time was Edward Hyams, who is credited with bringing vines back to Britain after finding Roman vine remains nearby. He is also another one of that rare breed who believes our wealth depends on the soil and labour. In *Soil and Civilisation*,[18] he said the first stage of civilisation was exploitation of the soil. He showed that 'the crafts of tillage (of wheat) do not develop in Europe slowly before the eyes of the archaeologist as they do in Mesopotamia' (where wheat was first cultivated). In other words, plant crops don't move around all on their own, but labour skills are an essential part of the movement in growing the crops elsewhere. The great agricultural scientist and socialist N.I. Vavilov also recognised this. When collecting a quarter of a million different sorts of seeds from all over the world, he also learnt over 20 languages to find out how they were cultivated, as this was essential in their growth when transplanted to his seed bank in St Petersburg. He developed the theory that there were eight centres of origin of all cultivated plants.[19]

Usually crops moved across the world from small farm production to plantations. Look at tea from China into East Africa. Rubber moved from the Amazon, where collection was haphazard, to rubber plantations throughout Malaysia and Indonesia where owners could guarantee production – by moving Tamil workers from India. Then there is the biggest crop migration of all – soya. Shiploads are being exported from massive soya farms in Brazil to China, where it was first cultivated. As shown in Chapter 3, bananas were first cultivated in Southern China, but are now grown in the Caribbean. Sugar cane probably evolved in New Guinea, and the variety called 'Creole', now grown in the Caribbean, is almost identical to the varieties there. We know 'Creole' as a simple language[20] that evolved for American slaves to talk with each other on the plantations, without their owners understanding.

Unions always find it difficult to recruit on plantations. That is why it is such a powerful mode of capitalist production. Unite tried over many months to recruit in one big operation. The management agreed to allow the workers their entitlement to holidays, but when they came back from holiday – there was no job. Security on some plantations is like a prison camp with patrols constantly circling the perimeter and patrolling the corridors, carrying out spot checks. I have heard workers calling them 'open prisons'.

Slaves were the first form of labour in plantations. However, there are modern forms of slavery still all over the world – look at the palm oil industry,[21] needed to make our ice cream. A Home Office Report[22] said that there are estimated 10,000–13,000 slaves in Britain, supporting the need to introduce a Modern Slavery Bill.[23] Modern slavery can be defined as 'human trafficking' or 'forced labour', the latter most common in the food supply chains. The minister introducing the Bill, Karen Bradley, reckoned that nearly 20 per cent of all UK slaves are in the food and agriculture chain.[24] That is 2,000! The Joseph Rowntree Foundation Report on 'Forced Labour'[25] chronicles many of the goings on.

In June 2017 a Nottingham landlord was jailed for modern slavery. He forced two men to work – one on a farm – and while they earned up to

£400 a week, he gave them less then £30. A farm supervisor tipped off the Gangmasters and Labour Abuse Authority (GLAA – formerly the GLA).[26] The GLA was set up in the wake of the Chinese cockle-pickers tragedy in 2004. A former Director of GLA, Dr Ian Livsey, told me that they have probably made a good job of clearing out exploitation in the 'top tier' of food production, but find it much harder when labour is subcontracted over several tiers.[27] Since then the GLAA helped bring the case against the Nottingham landlord.[28]

According to my botanical bible:[29]

> More significant in making a plantation a distinctive type of (tropical) crop producing unit than mere size, origin of labour force or nationality of controlling interest is the way in which production is organised. It has already been suggested that a plantation bears more resemblance to a factory than a farm. It is an economic unit organised for the efficient production of a particular crop, and just as a factory is organised for the manufacture of one specialised item so a plantation is concerned with the output of one end-product … these practices require the possession of capital on a scale usually available only to companies …
>
> Of the technical advantages of the plantation, perhaps the most important is the fact that, thanks to its size, it can make full use of the benefits accruing from the division of labour ... the plantation not only employs specialists for all these jobs, aided by machinery where appropriate, but can call on the services of mechanical and electrical engineers, chemists, entomologists, pedologists (soil scientists), accountants and many other experts.

Eric Williams, Walter Rodney and Karl Marx all contended that the global capitalist economy was largely founded on the creation and produce of thousands of slave-labour camps based in colonial plantations exploiting tens of millions of abducted Africans and Asians. I wonder what Marx would have had to say now about the newer temperate version. It is more than just a fight in a few fields in Eastern England. It was the plantation owners in England who wanted rid of the AWB. Europe has similar modes of food production, undermining more sustainable forms of farming. At present the ILO Convention on Plantations[30] applies only to tropical crops. It would seem that it needs bringing up to date for temperate countries, bearing in mind the similarity in conditions of monocultures, migrant workers and modern slaves. Plantations once defined colonialism. Perhaps some social historians should see how temperate plantations define neoliberalism.

Agricultural Science Worker

It was while I was studying plant science, especially these plantations, that I began to consider land and labour more. While at Wye College, I began to see myself, not just as a 'scientist', but also a 'worker'. I was the 'head' part of 'by hand and by head'.

For the 7–8 years I studied various agricultural degrees, I was encouraged to believe in the ability of markets and science to solve all the food problems. On the science side we had to be objective, keeping our head down and not looking at the rest of the world. I never heard about workers. Together with the magic of the market, we could grow anything and everything. It was overlooked that somebody had to put the stuff in the ground, look after it and bring it to us. The workforce was invisible.

I heard about farmworkers when our Wye College Socialist Society invited Joan Maynard, then President of the National Union of Agricultural and Allied Workers (NUAAW) to speak. It created quite a stir. I remember being at a party, and one character saying: 'I'll tell you why farm workers are thick. They wouldn't be farm workers if they weren't. Ho, ho, ho.' Joan told us about tied cottages. I couldn't believe I had never heard that many farmworkers lived in accommodation that was 'tied' to their jobs. If they lost their job, they lost their cottage. They daren't complain about their working conditions, for fear of losing both their job and their home.

Being an agricultural science 'worker' kept me going through my PhD, as it was boring counting half a million specs of life. Most biologists end up counting – whether it's spots on leaves or spores on roots. I'd then imagine that it must be as boring working on the Austin production line. I kept going because of the prize of a degree at the end, but I nearly didn't. I was ready to put some flowers in my hair, but thought no, there is something more important here – soil life.

Since then I have been fascinated about the relationship between labour and the land. Plants can do a lot, but without people they just grow wild. We would not eat if somebody didn't put the stuff in the ground, look after it and harvest it. But who put that 'stuff' in, why and where, has shaped the society we are today. And will shape the future.

In Britain, we seem to treat our food and farmworkers particularly badly. I wonder whether that is because we can always threaten them with food from abroad. Elsewhere in the EU, they produce much more of their own food, so are more reliant on their rural populations. The French are proud of their farming community, and while they only account for 3 per cent of the population, French farmers have a lot more political clout than farmers in the UK.

Farm unionisation started with the Tolpuddle Martyrs, followed by the formation of the National Agricultural Labourers Union in Wellesbourne in 1872, by Joseph Arch, as a consequence of the decline in UK agriculture following the repeal of the Corn Laws. They have gone through many re-organisations, becoming the National Union of Agricultural and Allied Workers from 1906–82. In 1910 major strikes and disputes broke out in three Norfolk villages for 1 shilling a week extra. The NUAAW became part of the Transport and General Workers Union in 1982, which merged into Unite in 2007 as the Rural and Agricultural Workers National Sector.

Photo 7 Joseph Arch in 1872 at the time of the formation of the National Agricultural Labourers' Union.

Photo 8 Ivan Monckton and Steve Leniec from the Rural and Agricultural sector of Unite.

AGRICULTURAL WAGES BOARD (AWB)

Even Margaret Thatcher accepted the argument that farmworkers were in a difficult position to bargain, so she saved the Agricultural Wages Board when getting rid of the other Wages Boards. The AWB's main function was to agree decent wages between national representatives of the workers and the employers, to provide a decent and respectable return. Both farmers and farmworkers liked the AWB because it meant they didn't have to worry about bargaining. Many small farmers used the provisions of the AWB when they tendered their services to other farms. The AWB provided set agreements and rates without individual negotiation. The AWB also dealt with issues relating to tied cottages and structures for skills rewards.

Since the abolition of the AWB, the principal protection comes from the EU Working Time Directive, which sets standards on the length of the working day. That may go, as it reverts to UK control (see the report on the state of agricultural labour from the Food Research Commission[31]).

Farmworkers' wages were usually a few pence an hour above the Minimum Wage, which was introduced in the late 1990s by the Labour government. This was in part because farmworkers are geographically isolated from each other, so find it difficult to organise as other workers can. 'We can't have that' cried the plantation owners. 'We don't want to pay migrant workers more than the Minimum Wage!' More exactly, they said it wasn't just a matter of paying a few pence above the Minimum Wage, but complicated in that workers out in the field are covered by the AWB rates, but when they return to the packhouses are paid Minimum Wages. Packhouses are the indoor workplaces, where the fruit and vegetables are washed, graded and packed. I had the answer – pay them all the time at AWB rates.

In the early 2000s, moves were made to abolish the AWB, led by the NFU and Fresh Produce Consortium. The Lib Dems swore they were against the abolition, and signed pledges to that effect. Plans to abolish it were dropped. But then came the Coalition government in 2010. The Tory pledge was to have a bonfire of the quangos, which included the AWB. To the Lib Dems we asked: 'Why are you going along with abolition, as you have always been against it, as evidenced by a petition signed by your Agricultural Minister?'

When the Bill to abolish the AWB came to parliament in October 2011,[32] there was a period of consultation during which Unite managed a late lobby. In early 2012, there was some sort of Commons debate where the government's impact assessment stated that the abolition would mean about £250m would be removed from the rural economy. I went, with others in the labour and trade union movement, to lobby the then leader for Lib Dems, Tim Farron, at his home surgery in Cumbria. He almost laughed us out, saying he couldn't believe we wanted to keep something so antiquated, and that he hadn't heard any voices against abolition from local farmers.

The AWB was abolished without a proper vote in the Commons in late 2011. The Welsh Assembly had complained that the AWB applied to both England and Wales, and they did not want it abolished. They said that because agriculture was a devolved matter, the situation in Wales could not be decided in the UK parliament. To get round that, the government put the abolition into an Employment Bill, saying 'employment' was not a devolved matter. The abolition barely got any debate in the Commons.

The only decent discussion was in the House of Lords. Labour opposed an abolition amendment in this wider Bill. Several lordships turned up to say what a waste of space the AWB was. Most disappointing in this bunch was Lord Curry of the famous 'Curry Report' on 'Food and Farming – a Sustainable Future' in 2002. He is in charge of the Leckford Estate belonging to Waitrose. When Waitrose, the 'cuddly, locally-sourced, free-range, organically-produced, GM-free' supermarket (as *Private Eye* describes them) was asked about the abolition, they said 'we do not take any view' on the abolition. Lord Curry said the AWB was a 'relic' and could only see it as 'red tape'. He said that anybody who resists change must think agriculture was still in the *Lark Rise to Candleford* era, where employees are exploited by unscrupulous employers. I was to bump into Lord Curry when speaking after him at the opening of the City of Durham Sustainable Food Policy.[33] I said that the trouble with a lot of EU payments is they enable landowners to shoot grouse for nothing. He was livid. The chair had to keep us apart.

However, what was good to hear in the Lords' debate was Lord Greaves, previously the Lib Dem representative on DEFRA matters, and a character I had come across in local council politics. He said that AWB wasn't in the Lib Dem manifesto, so he could oppose it, adding the following:

> I received a message from somebody local ...The letter was sent to me by an acquaintance of mine in Pendle, Charlie Clutterbuck. He sums up many of the problems when he states:
>
> *'On this occasion I wanted to raise the issue of the abolition of the Agricultural Wages Board, and the general disappointment that the Lib Dems are going along with it. It surprises some of us and it seems a most surprising turn around.'*
>
> This is the point at which, like all my colleagues, I give people a lesson in coalition government – in trade-offs, compromises and all the rest, which I perfectly accept. There are times when an issue is not fundamental, and when we, as a party, should turn round and say no. Mr Clutterbuck continues:
>
> *'It is also odd that there are few farmers round here'* – this is upland Pennine Lancashire –*'who really want this abolition. It won't help them with their many other problems, by adding the difficulties of employment complexities. I was Chair of the Governors at Myerscough Agricultural College, and know of nobody who thinks this move is going to bring rural prosperity to areas like ours.'*

He goes on to say, in his words, not mine:

> It is clearly motivated and being pushed by the 'Plantation owners' in the East who hire hundreds of migrant workers and want to pay them a penny or two less an hour. 'Rural' won't benefit, but 'retailers'

> will. And in smashing up the AWB, they smash up the whole skills structure, on which most permanent farm workers depend for their career. I am on the LANTRA England Council and know that 'growing' skills are going to be crucial in the future. Doing away with the AWB Skills scheme sends the wrong message to anybody wanting a career in farming and who doesn't own a few hundred acres of their own land.[34]

Also opposing the abolition was Baroness Trumpington, previously Agriculture Minister in Mrs Thatcher's government. My farmworker mates bumped into her in the Lords and persuaded her to oppose the abolition. She bought them all a drink – and it wasn't coffee. The Bishop of Hereford, also opposed the abolition, quoting Churchill (1909), introducing the Wages Boards over 100 years ago, saying that this was as relevant today as it was then.

> It is a national evil that any class of Her Majesty's subjects should receive less than a living wage in return for their utmost exertions … where you have what we call sweated trades, you have no organisation, no parity of bargaining, the good employer is undercut by the bad and the bad by the worst; the worker, whose whole livelihood depends upon the Industry, is undersold by the worker who only takes up the trade as a second string … where these conditions prevail you have not a condition of progress, but a condition of progressive degeneration.[35]

Churchill was then a Liberal MP. It is ironic that a century later the Lib Dems lined up to attack agricultural workers. With the Libs Dems' support, the AWB was abolished by being included in an Employment Bill, rather than an Agricultural Bill. The Enterprise and Regulatory Reform Act gained royal assent in 2013, abolishing the AWB.[36]

This manoeuvre was intended to avoid a challenge by the Welsh Assembly, led by Labour, which charged that as the full title of the board was the England and Wales Agricultural Wages Board, the UK government had no right to determine whether Wales should have an AWB. The Supreme Court of the UK decided that Wales was in the right.[37] The Welsh AWB will be reconstituted in a slightly different form. Scotland and Northern Ireland have decided to continue with their AWBs.[38] Only England has no AWB: a victory for the plantation owners, over family farms and all farmworkers.

FARM FATALITIES

Since the AWB abolition, the principal protection for rural workers comes from the EU Working Time Directive, which sets standards on the length of the working day and the other EU-based Health and Safety Directives, popularly known as 'The Six Pack'. The important safeguards for workers are based on the UK's Safety Representatives and Safety Committee Regulations (1977). However, the NFU do not recognise the farmworkers' union so there are no recognised safety representatives on farms. Unite has tried to get 'Roving Reps', but with very limited success. The idea of safety committees on farms is impossible to imagine.

In the *Hazards* movement, consisting of many trade unionists coordinated through *Hazards* magazine, we argue that 'union organised workplaces are safer workplaces', and have the statistics to prove it. However, the reverse is true for farmworkers – poorly organised, they suffer. Farms are the most dangerous place to work in Britain, in terms of deaths at work. Almost as many farmworkers died in the killing fields of this country as UK soldiers in the Afghanistan War. In the 2000s, there were over 40 farm fatalities a year on average[39] (400 in ten years) compared to 454 British Forces' fatalities (2001–14).[40]

One of the NFU representatives at the HSE used to argue that farms under the Red Tractor scheme – whose logos you see on food products in supermarkets – were safer. Yet the scheme incorporates no health and safety standards in its certification. Farming and the construction industry used to be on a par with each other in terms of fatalities at work, but the building industry has reduced fatalities by half in the last 10 years, thanks to unions and companies working together. In farming, the NFU has virtually taken over from the HSE with its own 'Farm Safety Partnership'. The HSE does not carry out surprise inspections and there is no sign of any reduction in farm fatalities.

The awful statistics haven't really changed over the last 15 years – the last improvement came when 'roll bars' were introduced in the 1990s, saving many a worker when a tractor rolled. There is a prevailing culture of 'we've always done it that way'. As the average working age on farms is around 60 years, it is not surprising that most deaths are among older self-employed farmers and likely to continue. Probably the best way of saving lives would be to change the tax system to stop farmers working till they die in order to pass their farm to their children.

Over the past few years, I have worked hard to try to reduce these fatalities, suggesting to the HSE Committee to develop vocational qualifications for health and safety in agriculture and horticulture. With the help of five Awarding Bodies and two Sector Skills Councils, the HSE, Unite and the NFU developed a suite of health and safety qualifications at the three vocational levels – 2, 3 and 4.[41] Around 2,000 people undertake these qualifications each year, mainly level 2 in agricultural colleges. So the prospects are the culture will change. These qualification skills were built into the AWB in 2012, to encourage people to undertake the qualifications. I praise the NFU for agreeing to reward these accreditations. However, with the abolition of the AWB, that reward system has been lost.

The working conditions of farmworkers on organic farms, it seems, are no better than anybody else's, at least for migrant workers. The late lamented Gareth Edward-Jones, with whom I had the privilege to sit on the government's Advisory Committee on Pesticides, examined conditions for workers in organic and non-organic farms. The health of over 600 workers was measured through the use of four standard health instruments. While farmworkers' health was significantly poorer than national norms, there were no significant differences in the health status between conventional and organic farms. However, organic farmworkers scored higher on a fourth health instrument, indicating that they were happier than their counterparts on conventional farms. The organic certification organisation IFOAM[42] includes working conditions in its 'Principles of Health'. The authors of this research questioned whether their working conditions achieved these standards. I got the feeling they thought they hadn't.

I worked with the Pesticides Action Network UK and Unite to survey migrant workers about their experience with pesticides and raise awareness about the requirements of the EU Directive Sustainable Use of Pesticides. This requires member states to design training programmes 'to ensure that sufficient knowledge on the following subjects is acquired: 1. All relevant legislation regarding pesticides and their use: 2. Hazards and risks associated with pesticides and how to identify and control them, in particular their legal rights.'[43]

THEY'RE OFF!

Following the Referendum, the Environment, Food and Rural Affairs Committee (EFRA) Inquiry, 'Feeding the Nation: Labour Constraints', heard from a wide variety of witnesses representing various agricultural and horticultural employees. The report said businesses were unanimous in reporting they had long struggled to find sufficient labour to meet their needs, either from UK or overseas sources. They considered that these problems had worsened since June 2016 following the UK's decision to leave the EU, despite efforts to entice UK workers to their sector. 'Employers told us repeatedly that UK workers did not want the work.'[44] I'm not surprised. The decline in numbers is due in part to the fall in the value of the pound, making work less worthwhile, but also to the feeling of not being wanted.

The same Inquiry put the findings to ministers, who thought they were exaggerated and that labour supplies would be fine. The ministers talked about bringing in apprenticeships, increasing skills and qualifications, encouraging a worthwhile career and building positive perceptions of the industry. I would love that, but there has to be some long-term injection of money. Otherwise, I can't see any of this sort of development on these wages – remember they are down to minimum levels thanks to the abolition of the AWB! Oh – the government also plans to 'reform the benefits system aimed at encouraging people back into agriculture'.[45] I've heard two suggestions – one that there should be 3-month conscription, like in Switzerland, for youngsters to work in the fields, and the other is for prisoners to do the work. It all sounds a bit desperate.

I would agree with the ministers on the importance of apprenticeships, skills, qualifications and careers. This is all very well, but it is not just a wish list. There has to be serious long-term investment if young people are to make a career in rural areas. I remember when I was a rare Labour Chair of the governors at Myerscough, the Lancashire College of Agriculture; there was a dreadful downturn in the number of young people wanting to study agriculture and horticulture in the early 1990s. There are signs of an upturn, but there have to be long-term prospects for young people to set up businesses in the countryside that employ local workers on a living wage. For many small businesses that may be off-putting. However in food and farming, if we used the present subsidies in the way I suggest, wages could be subsidised up to the living wage. There would be more than enough money if we transferred subsidies from large landowners to farmworkers in the form of a living wage.

The death knell for land-based skills development came during the Coalition government 2010–16, while I was on the England Council of LANTRA – the body responsible for developing skills on the land. Up till then, the land sector had a distinct slice of the skills cake. However the government decided land skills had to bid with others for a piece of the action. I knew we could not compete. And so it proved to be.

The potential reduction in migrant labour post-Brexit may catalyse structural change in the agricultural and horticultural sectors. The Agricultural and Horticultural Development Board (AHDB) looked at the impact of a UK exit from the EU on the agricultural labour force and analysed the reasons why the UK agricultural industry has increasingly drawn on EU migrant labour and what effect any future restrictions on the free movement of people might have.[46]

The availability of overseas labour could be mitigated by an increase in productivity through innovation and skills development. Restrictions in the availability of labour are likely to increase the costs of employing staff. This is said to be happening already and could lead to investment in more capital-intensive production systems such as automation. However, businesses might find that risky, as labour shortages could make them less competitive in the global marketplace. Not all parts of the plant and animal sectors lend themselves to automation. There may be substitution of capital for labour, but because of uncertainty over Brexit and trade deals, this sort of investment may be put off. 'We could see both the current structure of the industry and the nature of UK agricultural production change significantly as a result.'[47]

A 'hard' exit – where tariffs appear in all their glory, together with a lack of alternative labour schemes – would leave a large section of the industry vulnerable in the short term. Without support for investment, and possibly unable to employ EU and other migrant workers, the UK agricultural industry will become less competitive. No wonder people in the plantation plot are crying for tariff-free access to the Single Market. But the cake has been eaten.

According to the Senior Executive Officer for G's (Fresh), who grow 60 per cent of the UK's celery, radish, spring onion and beetroot, and employ 4,000 workers in this country:

> If we can't bring people to the work, we can take our work to them … We rely on our EU workers, who are seasonal workers, whereas local people want permanent jobs. Seasonal workers go back home each year, so don't impact on immigration figures, they don't bring their families, don't impact on social service and they pay substantial amounts of tax. If it was stopped, it would shut down our industry.

> We were promised a scheme like the old SAWS scheme by senior politicians, but that has gone quiet.[48]

Technofix

Some engineers are suggesting there are technical ways to reduce the need for migrant workers. This is a big issue for the future. Many consider that there must be less reliance on migrant workers, and the classic way to deal with that is to introduce technologies to replace the manual parts.

One carrot farm in Yorkshire has already taken advantage of technology to overcome any labour shortage issues. Machines have replaced 18 migrant workers at the farm, using electronics to measure the size and health of the crop.

The think-tank Resolution Foundation report[49] says that such changes could potentially offer new opportunities for automation to counter economic disruption. Author Adam Corlett explained: 'Looking at those sectors with the highest proportion of EU migrants, we find that some … face relatively low prospects for automation, while others – such as agriculture, food manufacturing and food and drink services – may see new pressures or opportunities to automate.' The Foundation believes that Brexit could give a push towards increased 'productivity' – i.e. more productive labour.

There are difficulties with increasing automation in the agricultural industry. The use of technology has proved challenging in the harvest of fragile soft fruits and vegetables, but it is likely that there will be a gradual move towards automation. It is a shame we closed the Silsoe Institute of Agricultural Engineering Research Station about 10 years ago.[50] This would have been the place to develop this sort of engineering. We need smaller-scale machines, capable of doing different jobs. We rely for research on foreign companies like John Deere, a US company, which favours large machines. Other machine companies include New Holland (Dutch), Massey Ferguson (USA), Case IH (Italy) and Claas (Germany), all of them more concerned with big machines and productivity. We only have JCB.

In 2015, the government introduced regulations allowing larger and faster tractors and trailers.[51] The maximum speed is supposed to be 25 mph, but tractors as big as double-decker buses now hurtle through our village. Remember the high rate of farm fatalities is due to moving animals and machinery. I was on the HSE's Agricultural Industry

Advisory Committee when this regulation was introduced, and asked for a 'risk assessment'. I never got one. I predict an increase in tractor accidents on roads, especially with trailers – which are not governed by the same regulations as lorry trailers.

Investment

There is reluctance to invest because of uncertainty about what Brexit means. Producers may be reluctant to increase capital investment at the level required to offset a potential loss of labour supply. Interest rates are at an all-time low, but reassurances would be needed from government for their future policy – do they want to export more booze abroad and import more cheap food *or* put in place ways to make British farming a great career and an attractive long-term investment?

There is also the matter of Quantitative Easing (QE), as we saw in Chapter 1. One of its continuing consequences is that in order to fund pension schemes, businesses cannot risk investment in capital projects.[52] QE was supposed to inject money into businesses, but the reverse seems to be the case.

The AHDB argues that the potential restriction of migrant labour in seasonal agriculture may be understood as a 'catalyst' that forces the industry to improve productivity.[53] But businesses will be asking: can they replace this migrant labour with other migrant workers?; can they their hold wages down?; who else could they employ – students?; how risky is any capital investment?; where is the research to help?

In the long-running BBC radio soap *The Archers*, Alice Archer, a bit of a techie, explains to Adam, who wants to improve his soil carbon, that drones could fly over his 'mob grazing' plots, taking photos of the plants below. Using some clever software, these photos could be analysed to see if he was getting the right mix of plants, and thereby rectify the situation if there wasn't enough diversity. We can thank Graham Harvey, the agricultural plot adviser, for showing us how high tech and sustainability (see Chapter 7) can be compatible. This raises the question of who owns the information produced. I suspect it will end up in an information databank held by one of the big companies, gaining ever more power and control, whereas it would be preferable if it was part of a publicly-funded research unit that could pass on the information to other growers, free of charge.

Wider afield, we hear of robots taking over our jobs, as in the milking parlour in the Archers. I can see driverless tractors soon, with accurate GPS guidance systems. I can see drones spraying pest outbreaks. But I don't see robots replacing migrant workers just yet. The history of technology innovation at work is that it first replaces expensive labour. The car-body sprayers were replaced first in our local Birmingham car factories – not the packers. Women were always left packing the biscuits, while the rest of the process had been automated. 'It's because women have nimble fingers.' No! They cost less! In farming, it will be expensive drivers and agronomists threatened. If the work can be done by cheaper labour, as has been the case in the food-service sector, it will be unaffected. I don't see robot waitresses just yet.

The joy of food is it cannot be replaced by virtual food. The completely natural, unalterable property of our eating is that we need real food. It will always be real. Real vitamins, fibre and other 'goodness', as my mother would say. It can never be replaced. The food and farm sectors will go on, when others go down. However, parts of food production can become virtual, and web-based technologies will replace some jobs in the food chain. Virtual jobs could replace real jobs in several food-related areas, such as logistics. Where once we saw vans with food manufacturers' names on, now we see dozens of logistics vans. You can't help but think the suppliers and providers could link directly with clever apps rather than centralised logistics. Home deliveries could be developed countrywide by food retailers. We order our takeaways on a 'Just Eat' app – although we may still need a poorly-paid Deliveroo worker to deliver.

We know all about quality systems, hygiene systems, health and safety systems such as HACCP, and environmental standards. Michael Heseltine promoted these standards, many of which were created by the British Standards Institute, as one of our great exports. We could transmit these standards electronically. HACCP[54] enables 'food business operators [to] look at how they handle food and introduces procedures to make sure the food produced is safe to eat'. It helps food businesses comply with the EC Food Hygiene Directive (93/43/EEC) based on the principles laid out by the *CODEX Alimentarius* Commission – the body which sets out WHO and FAO standards.[55]

We could train people in Africa and Asia, using the web, to cope with these systems and deliver good-quality, healthy food. A government official in Zimbabwe reminded me that many developing countries see these systems and standards as barriers to accessing various markets. Yet the WTO says they are not barriers to trade. They aren't if you are a corporate. Otherwise they are daunting. We could deliver the training required to meet those standards and systems over the web.

In other sectors, new internet technologies have enabled the likes of Dyson to manufacture in Malaysia, with local labour. Dyson provided the required training from the UK using 'Skype'-like web connections, with software programmes entering data at the other end. There must be a parallel in food. Rather than just moving foodstuffs about, we could help develop African and Asian food sectors. We could build food businesses overseas, as we have done here, where our food sector is ten times larger than our farm sector. We should not just treat former export markets as trading partners, extracting their resources. We can do more than that, and encourage a business-centred and entrepreneurial approach to developing food businesses overseas – something we are good at – based on their local foodstuffs.

Technology and Labour

The relationship between labour and food production technologies is fascinating. On behalf of the British Bakery Workers Union, I headed up a leg of a tripartite EU course for trade unions in the food industry looking into possible technical threats to their jobs. The three groups of trade unions met up for three weeks, one representing biotechnology union officials in Italy, another beer workers from Ireland and the third, bread workers from the UK. We were looking at factories using biological means of production. Bread is made with living yeast. Well, that was until the Chorleywood Process did away with that. The phrase 'greatest thing since sliced bread' gets its reputation from this process – it uses more British flour, and replaces the slow mixing and kneading process, taking an hour less to make.[56] About three-quarters of all the bread we eat is made using this process. Beer has kept the yeast, only with more water. Yogurt, too, requires living yeast to make it work. Together they are called 'biotechnologies' – not to be confused with genetically modified organisms.

Each leg of the course showed off the local culture. While we were based in the fabulous GMB Education Centre in Manchester, we went off for fish and chips in Blackpool and a sing-song in Liverpool. We went to local Rank Hovis McDougall bread factory in Wigan. I'd heard from a bakery worker that you can tell when the white stuff is fully cooked, by putting your hand into the loaf and grabbing a palm full, rolling it round and dropping it on the floor. If it bounces back to half the height, it's ready. The Italian union reps were horrified, not just by the bread, but by the hours these workers were working – shifts covering 24 hours a day. They threw up their hands shouting 'Compulsory overtime, Non, non, non.'

Then we went to Ireland to enjoy their culture and biotechnologies, which meant a trip to Guinness. We went round another 24-hour process, where the workers again worked shifts. And again, the Italian union trade union officials said: 'You shouldn't be letting this happen!' They kept asking 'Can you not make beer or bread in less time than 24 hours?'

So we were all curious to see what happened on our visit to Italy. We went on a trip to the yogurt factory, eager to see what they did. They showed us where the milk came in and where they put it into vats, and turned it on for a few hours. They came back and packed the yogurt in containers, ready for dispatch. So one of the Irish officers said – 'Why do you ferment the yogurt for 3.5 hours. Couldn't you make it either shorter or – better – into a 24-hour process?' The Italian response was: 'We want it this length of brewing so we can come to work, put it on, then pack it up, and then go home to our life and families.' They taught us a lesson, as they saw that the technology should be in our control; we should not be at the behest of the technology. There won't be any chance to learn these cultural lessons now we have left the EU!

SUBSIDISE LAND WORKERS NOT LANDOWNERS

The central thrust of this book is that the best way to spend £3bn of the old CAP subsidies is not to give it to landowners but to land workers. That money could work for us. We could produce more food and all manner of useful services.

Any accountant will tell you the most costly element in any work is labour. So subsidise it! The main cost of any food is not the raw ingredients, but all the processes between plough and plate.[57] We need to keep those costs down, while at the same time paying decent wages. At present we are caught between customers wanting cheap food and producers, via the retailers, having to keep labour costs low in order to do so.

Subsidising wages would encourage more local people to see better prospects in working on the land. They could have apprenticeships, make careers, or get permanent well-paid work on the land. Instead of subsidies going to the owners of land they would go to workers, where it will do a lot more good, not only for the workers, but also for employers and customers. If it costs employers less to pay workers decent money, it keeps food costs low, as the saving could be passed on to the consumer, thus benefitting everyone in the food chain.

In order to receive the subsidy, employers would have to provide decent working conditions. We should reintroduce a form of the now abolished AWB. The devolved nations have something like the old board already. It would only be England that has a job to do. These boards would do what they did in the past – pay decent wages, reward skills development and encourage improved health and safety. All farm employers would have to recognise the new Farm Boards and safety representatives.

Subsidies could be directed to those areas of work that produce healthier food – both for people and the environment. In particular, the growth of foodstuffs that capture carbon in the soil should be rewarded, as we will see in Chapter 6. By subsidising these foodstuffs, it would encourage the rest of the food industry to use healthy and sustainable ingredients. If local food is healthier and cheaper, we are on the way to creating a revitalised British brand – coloured red for labour and green for the environment.

For instance, most agree with 5-a-day diet of fruit and vegetables, so we could – easily – produce more fruit and vegetables locally. They grow well in our climate. The best way to do that would be to subsidise the workers who pick and pack. We should also reward mixed farming, organic and pasture-fed animal rearing. With these funds, they could diversify – to trees, bushes and fruits. We need diversification so we don't saturate any one market.

The Small Tenants Association complains that the CAP subsidies – for land – don't get passed on to them as tenants. So the new scheme should find a way to pay working farmers. This would help sheep and dairy farmers, and replace the subsidies they get now. Anybody working the land should get support. If the Duke of Westminster can be seen working, he too should be entitled to £10,000 – but not the £400,000 he gets now.

When I say 'subsidise land workers', I also include 'land science workers'. We have to revitalise our land research stations, to start asking and answering a much wider range of questions. We could use some subsidies to support the work of land researchers looking into more fruit and vegetable varieties. Our food is not just about 'food security' – the buzz word in most fund applications – but also about resilience, diversity, soil systems, carbon-friendly techniques, integrated pest management, local seed banks and agro-ecology generally.

£10,000 a year for 300,000 people is the sort of sum that could encourage people back to the land, in new and exciting ways. It would also build local economies and fits with new ideas for a 'Universal Basic Income' … we could call it the Basic Rural Income.

'Basic Rural Income' could enable enterprising farmers to diversify. There is much talk about this, but little real support. A basic rural income would encourage somebody to have a go, knowing that they had a fall-back, giving people more chance to try something new. We are going to need a thriving rural farm/food business if we are to develop our food and farm economy in novel ways and reduce that $66bn bill we pay for our food imports.

How could these funds be managed? It is clear the Rural Payments Agency is not up to the task. They could not even distribute the Basic Payment Scheme, which is pretty basic. They were fined – by the EU – for performing so poorly. I propose a possible solution in the final chapter.

Direct subsidies completely oppose free-trade mantras. However, they would start to address the major challenge we face. Most people want cheap food but cheap food leads to cheap labour and costs the earth – as we will see in the next chapter. By paying people to look after the land while producing our food, we could have cheap food, decent labour and better earth. Directed subsidies could also lead to saving a lot of food imports, building in resilience, so that whatever deals are struck round the world, our rural economy thrives.

That Marx character wrote: 'Capitalist production develops technology, and the combining together of various processes into a social whole, only by sapping the original sources of all wealth – the soil and the labourer.'[58]

So let's turn now to that soil – the land.

6
Land

In this chapter we look at the land. There are all sorts of way of working the land, producing all sorts of different products and services. We cannot directly control what happens, as it is not 'our' land. However, if we are to have more control over 'our border', perhaps we should argue for more control over our land.

The land can be viewed as lumps of soil. Dirt to many people. The soil is the top layer of the Earth's crust and is a complex living entity, consisting of minerals, water, air and living organisms. We call our planet 'Earth' because of the soil. Yet, for nine-tenths of the life of this planet, there was no soil. Since its creation, most of the natural world we know depends on it. This is probably the last great kingdom to be discovered. We have much to learn about how the soil works, as we'll see later in this chapter.

One of my reasons for writing this book was because I feel like a 'shop steward' – in union language – looking after the soil. We need to pay much more attention to our soil, now we are proposing to control our borders more. We have a real opportunity to come up with a soil policy that carries some weight. But the signs are not good – and this is one of the reasons I voted to Remain. The clues were there when the NFU opposed the EU Soil Framework Directive, saying it wasn't needed, and Hilary Benn agreed. They helped form a 'blocking minority', along with France, Germany, the Netherlands and Austria, mainly because the directive required a soil status report when somebody sold land, to make sure it was not contaminated. Together they managed to halt the directive in 2007[1] and it was officially withdrawn by the EU in 2014.[2]

However, we cannot decide what to do with the soil, because that decision is not in our hands. We do not control our green and pleasant land. The Crown does. The Crown owns virtually all land and then grants away most of it either as freehold or leasehold. These concepts of tenureship go back to feudal times – 1066 in fact. It means we are all tenants of the Queen on the basis of feudal superiority.

Few of us 'own' very much land. You need 10 acres to collect EU funds. Most of us are pleased to have a few square metres of garden to play around with. Yet urban dwellers *pay* around £35bn in various land related taxes, while rural landowners *receive* more than £3bn.[3]

There are about 60 million acres of land in the UK. An acre is about half a football pitch, so every woman, man and child in the UK has a kick-around area of just less than half a pitch. The rate of urbanisation is around 15,000 acres a year, and fears are often raised about concretisation, 'crowded' islands and land prices for new homes because of land scarcity. Yet for every acre of urban land there are nine rural – just look out of the window on your next train journey. Similar fears were expressed after the Second World War, when many new homes were 'taking over good agricultural land' leading to concerns about whether we could feed ourselves. The concerns were put to rest by colleagues from Wye College in their booklet, *Garden Controversy*.[4] They showed that, while we don't grow much grain in our gardens, nevertheless the output from gardens in new housing areas was every bit as much as from farmland – and that included the houses.

The landed gentry lord it over our land. Until 1999, all dukes sat in the House of Lords – if they bothered to attend, that is. Scrapping this hereditary right was Tony Blair's most radical act as Prime Minister. The Dukes of Buccleuch, Westminster, Cornwall, Northumberland and Beaufort, and the Queen, all own over 100,000 acres each.[5] In Lancashire, the same families have owned the land for 400 years. It is not merely the fact they own the land, it is what they do with it. The main activity on many moors is shooting grouse. While employing a few rearers and beaters, this land could be doing a lot more for local economies. If it were forested or properly conserved, all sorts of other jobs and businesses, tourism, for example, could bring money into rural areas.

The distribution of ownership of land in the UK is more unequal than the distribution of wealth. A mere 7 per cent of the population own 84 per cent of the wealth. Of the 60 million acres in the UK, 69 per cent is owned by 0.6 per cent of the population, giving Britain the most unequal concentration of landownership in the EU, bar Spain.[6] Half of Scotland is owned by just 432 landlords.

What this means in practice is that it makes it very difficult to work together for the better good. We can all come together when discussing pollution in the air, because air is not owned. Not yet. It means any suggestions for international, national, regional agreement can't work.

Whatever we may think is a good idea for the soil, there are a lot of powerful obstacles, many of them called Lords, in the way.

The landed gentry will be a force to be reckoned with. Many are (still) members of the House of Lords. Many will be receiving substantial EU subsidies. The Scottish National Party proposed a Land Reform (Scotland) Act that includes new protections for tenants, an end to tax relief for sporting estates and a new Scottish Land Fund with £10m available to help community buy-outs. A more radical proposal to limit the amount of land anybody could own wasn't supported by the SNP.

Land reform used to be part of the left's agenda. Alfred Russel Wallace jointly came up with the theory of evolution with Charles Darwin. He also founded the Land Nationalisation Society. In *Land Nationalisation*[7] he spelt out how labourers toil with the soil, but the resulting profit goes to the owners – for doing nothing. Lloyd George and the Liberals – then the Labour Party – championed land reform. After the Second World War, the Labour manifesto promised to work towards land nationalisation, with fair compensation and revenue for public funds from 'land betterment'. There have been recent calls for a land value tax. As we take more control over our borders, we have a chance to take more control of our land, and perhaps receive funds for public goods when land is 'bettered'.

NOT ALL LAND IS THE SAME

There are five main categories of land in the UK, classified according to international criteria. Grade one land is ideal for arable farming and can thus produce cabbages and brassicas. Grade two is nearly as good, but not as flexible to crop. Grade three is moderately good although it produces a narrower range of crops, but still can grow grain crops. Grade four is mainly pasture with forage crops, while grade five is pretty poor, good only for permanent pasture and rough grazing. We have more 'rough grazing' in Britain than the whole of Europe. This may be just down to geography, or lack of investment. These moorlands are seen as fit only for grouse-shooting, although we could do more for global warming simply by afforesting (reforesting) them.

But you can't grow vegetables on moorland. An economist called Ricardo said we should only grow whatever grows best in any place. We shouldn't grow 'marginally'. Yet many world crops are grown in less than optimum areas, which are nevertheless well-suited for machines or labour. Look at coffee grown in Brazil, which is often hit by frosts. Maize is pushing further north. There are other reasons for crops growing where they do – and labour costs are usually crucial. So it is not necessarily 'natural' that any crop is grown where it is – although there are limits.

UK farming is not homogeneous. There are two very different forms of farming in England. The big divide is either side of the backbone of England. To the East are the large arable plantations. They are the veggie growers, although nothing like your back-garden veggie grower. Miles of fields with massive machines the size of a house carrying migrant workers over a monoculture of crops. It is bad for land, as that is where most soil erosion is occurring.

To the West, there are the rolling fields of pasture – which Sir George Stapledon believed was so good for the land.[8] Here sheep graze, and cows can be still be seen, although more and more are getting locked up for longer periods. In the name of efficiency, dairy farms are becoming more 'intensive', relying less on grass, and more on bought-in feed – particularly maize grown in the East and soya from Brazil. In my own area I can see that in the last few years, many more barns have been erected, to house cattle for longer periods. These family farms on pasture land have twice as many soil creatures as arable land.

The difference between England, Wales and Scotland is that England has 15 per cent of 'less-favourable' land, whereas Scotland has 85 per cent and Wales around 80 per cent. Under Article 31 of the European Union Rural Development Regulations, people farming 'less-favoured' land are eligible for an annual income support payment. That allows farmers and crofters to continue to run viable businesses, avoids the risk of land abandonment, helps maintain the countryside by ensuring continued agricultural land use, and maintains and promotes sustainable farming systems. We shall see how the UK government might propose to replace this provision.

Northern Ireland is predominantly dairy with milk and cattle valued at over £800m, while crops and horticulture are worth around £150m. Their smaller farms are more 'mixed', meaning they grow both vegetables and grain as well as rear animals. They are good for each other – local crops for the animals are repaid by the animals fertilising the soil.

But it is not just our land we need to take into account. If we look at our food footprint, we see 70 per cent of the land used to grow of our food is located overseas, as we shall see in Chapter 7. Soya beans, used for animal feed, along with cocoa and wheat, account for most of that foreign land.

AGRI-ENVIRONMENT SCHEMES

About 10 per cent of CAP subsidies go to 'agri-environment' schemes, introduced in the early 1990s, whereby farmers are paid to 'green' their practices. Farmers volunteer to adopt different levels of schemes decided by the government for a period of five years. Examples of 'Environmental Stewardship', as the scheme is also called, include management of low-intensity pastures, integrated farm management, organic agriculture, preservation of landscape and historical features such as hedgerows, ditches and woods, and the conservation of high-value habitats and their associated biodiversity.[9]

A Natural England report assessing Environmental Stewardship schemes says: 'The analysis showed that, at the national scale, the density of walls and hedges within Environmental Stewardship agreements was not significantly different from the density of these features in the wider countryside.' There were also significantly fewer ponds and in-field trees under the scheme than in the wider countryside.[10]

A report from the University of Leeds says that these schemes are not leading to increased green awareness among farmers.[11] Even the most conservation-minded farmers have found the paperwork cumbersome and the rewards not worth bothering with. Nor do these schemes stop practices like cutting grass for silage rather than hay, which most farmers have turned to, as it is easier and safer. On our farm, we took hay late in the season. It was always touch and go whether we could get the hay in before the rain. Damp hay leads to farmer's lung – fungal spores that cause an allergic reaction in the lungs. A similar disease, *Aspergillosis*, is affecting workers handling organic compost. Silage is more reliable to cut, but it is cut earlier. It means there is a dramatic decline in the numbers of birds like lapwings and curlews, which rely on the grass seeds and insects that feed on them.

If we are serious about our environment, we cannot rely on individual farmers. We must work together to improve matters. The Game and Wildlife Conservation Trust suggest 'clusters of farms' working together. That would be a start; however, I would suggest that grants or subsidies should go to local councils to decide priorities, as part of their planning processes. We shouldn't be fighting over the agri-environment schemes, which were always a bit of a sop to the big European NGOs.

Agri-environment schemes are part of the CAP funds known as Pillar 2, and account for about 10 per cent of the subsidies. They do not begin to address the main environmental impacts – in terms of global warming potential. A whole new system is needed to deal with this, as we will see in the next chapter. There should be separate conservation funds for lakes, wildlife reserves, special scientific areas and to compensate the farmers affected. Regional funds could encourage groups of farmers to get together, to come out of dairy altogether, for example, to set up fishing/boating lakes, communal growing areas, glamping facilities and play areas. We will pick up on this in Chapter 12.

FORESTRY

Trees are a valuable resource. Not only do trees capture carbon dioxide and store carbon during the tree's life, they harbour four times the numbers of soil animals as arable soils. In the past, the UK was more forested. Only 13 per cent of our land is forested now (30 per cent public, 70 per cent private) providing jobs for 40,000 workers. This is the third lowest percentage of forested land in Europe (in the Netherlands and Ireland, the percentage is slightly lower). The Coalition government in 2010 talked about privatising half of the Forestry Commission but a campaign, begun by the forestry workers in Unite, put a stop to that.

Then there is flooding. Our village was flooded in December 2015. We need to plant many more trees to hold the water on the hills; even planting willows on the village green to hold water makes a difference. The National Trust caused an outcry when they purchased a farm in Cumbria, out-bidding local farmers. Melvyn Bragg and many local farmers were outraged. So was I. The Trust said they did it for conservation and flooding reasons. Nick Cohn in the *Observer* commended them. The *Observer* published my letter in response, which said Nick and the National Trust were right to be worried about conservation in the hills, but were hitting the wrong target when it came to who was responsible. It is not small farmers – it is the landowners who own vast tracts of land higher up the hills – the moors. I finished the letter with 'the trouble is, it is easier to outbid a small farmer than take on the landed gentry'.[12]

In Hebden Bridge, flooded three times in as many years, there is a 'Ban the Burn' campaign to stop the burning of heather on the moors.

Gamekeepers like to burn the heather to encourage young heather shoots for grouse, but locals say they are bearing the cost of this practice. The Royal Society for Protection of Birds (RSPB), along with Ban the Burn, lodged a complaint about the practice with the European Commission (EC), despite Natural England giving the owner consent to burn. This and other moors are supposed to be protected as 'natural sites' of special scientific interest and conservation, and if they were properly managed would reduce flood risk. In April 2016 the European Commission took the first steps in legal infraction against the UK government in relation to the burning of blanket bog in Special Areas of Conservation (SACs) in all of northern England, saying our government had not carried out a risk assessment as required by the Habitats Directive.[13] In April 2017, the EC sent a final warning, to which the government was supposed to respond within two months.[14] This is the stage referred to as 'reasoned opinion' and requires a response before the case is bought before the Court of Justice of the EU.[15]

Trees play a vital role in preventing flooding. In the 1990s a group of visionary farmers at Pontbren, Powys in Wales, near the headwaters of the River Severn, realised that the usual hill-farming strategy – loading the land with more sheep, grubbing up the trees and hedges, digging more drains – wasn't working. So they planted shelter belts of trees along the contours, stopped draining the wet land and built ponds instead. A consultant a few years later noticed that water wasn't flashing off their land, as it was nearby, and set up a research programme. The research came up with astonishing results, showing that water sinks under trees at nearly 70 times the rate it sinks into grass. The roots provide channels deep into the ground, so they act like a giant sponge. The research estimated that even though only 5 per cent has been reforested in the area, if all farmers in the catchment area did the same, flooding peaks downstream would be reduced by nearly 30 per cent. For many residents ravaged by Severn floods, that would make a world of difference.[16]

The Read Report's (*Combating Climate Change*, 2009)[17] key finding states: 'UK forests and trees have the potential to play an important role in the nation's response to the challenge of the changing climate. Substantial responses from the UK forestry sector will contribute both to mitigation by abatement of greenhouse emissions and to adaptation, so ensuring that the multiple benefits of sustainable forestry continue to be provided in the UK.' In 2011 the new United Kingdom Forestry Standards Guideline document, *Forests and Climate Change,* was introduced.

Forestry Commission Research Note 201, 'Climate Change: Impacts and Adaptation in England's Woodlands' describes the likely changes in climate, its impacts on forests, and offers factors that forest managers should consider when planning to make their woodlands resilient to climate change.[18] While there are strategies relevant to forestry and global warming,[19] it all seems to be about strategy rather than planting.[20] The trouble is the Forestry Commissions (FC) in England and Wales are not planting many trees. Four years ago, the FC was transferred to a new body – Natural Resource Wales (NRW), ending nearly 94 years as a UK organisation. The NRW planted no trees in 2016. You can say the Commission has reversed the decline of forests and turned it round by about 5 per cent from 90 years ago. But that is about it. Much more needs to be done.

In Scotland it is very different. The Scottish government's 'Rationale for Woodland Expansion' sets out a number of woodland creation priorities for Scotland that could deliver benefits for: climate change mitigation; economic and rural development; biodiversity and the wider environment; and community and urban regeneration. The rationale also sets out the wider land-use challenges of woodland expansion.[21] Part of the rationale is to establish a robust carbon sequestration monitoring and reporting framework.

Carbon offsetting through UK forestry can be of value when used in conjunction with strategies for reducing emissions. A study carried out at Kielder Forest calculated that the forest's 150 million trees lock up 82,000 tonnes of carbon annually. Roughly, each tree at Kielder is locking up 0.546kg of carbon per year – equivalent to 2kg of carbon dioxide. UK forests and woodlands contain around 150 million tonnes of carbon in the biomass and 640 million tonnes of carbon in the soil. UK forests and woodlands are a carbon sink, as they remove about 10 million tonnes of carbon from the atmosphere every year. In 2008, UK emissions of carbon dioxide were about 530 million tonnes per year.[22] It seems they count 'dead' carbon (burning a lump of soil), but not carbon molecules in live soil animals.

Brexit could offer some opportunities to increase the £2bn forestry sector with new plantings and restocking, according to the Confederation of Forest Industries.[23] They suggest fast-tracking UK forest standard schemes to replace EU-wide 'mandated characteristics' for certain construction products, and linking tree-planting with house development. They say there is an urgent need to integrate land-use policies involving forestry, farming and environment. Again, this is a local council issue, rather than one to be dealt with by individuals. The confederation says there should be transparent decision-making about what is most appropriate – say sheep or forest – and public funding of land-use management should be in the public good. Talking of which …

LAND USE

Changes to land use contribute more to global warming than almost any other factor, second only to the power industry, according to the Stern Report.[24] Whenever you look at the land, don't assume that it is only capable of being used as it is. The Incredible Edible Farm in Lumbutts, part of the Todmorden Incredible Edibles, is a few acres of what was marshland. Rubbish land in a valley in the middle of the Pennines, producing salad crops nearly all year round, several thousand pounds worth of fruit-tree cuttings, and supporting bees, ducks and cows. It has all been done without chemicals, according to the principles of 'Permaculture'. This is where ecologically-informed design arranges how plants grow together, and how to make the best use of the sun, soil and water. Above all, it shows that there is no problem producing food for the world, and for us.

A more famous example is the Eden project in Cornwall, where trees reach a height of 20–30 metres under great transparent domes. These 'biomes' are the world's largest conservatories, built in a sterile old Cornish clay pit, They are a living theatre of plants and people and their interdependence, showing and exploring possible futures. It reminds us of those Victorian masterpieces, like Kew Gardens and Chatsworth House, that were at the centre of our botanical Empire. We should keep these monuments to our creativity. They all show what we are capable of creating, when we have the will and the investment. They give us a vision of the sorts of things we could be growing in this country.

In the last few years we have seen the rise of solar farms and an increase in plastic cloches and polytunnels. Some people think they look unsightly, but they prolong growing seasons by 6 weeks. I'm more concerned about their working conditions. They are providing a wider variety of crops to feed our food service sector. Imagine what more we could do mixing these technologies.

The great crime I notice is the waste of land. I don't mean properly-conserved meadows, but the land just doing nothing, full of rushes

(*Juncus sp.*) and buttercups. I've watched the deterioration of much land in my part of the world in the last ten years. I say this just now as I have run several miles over fields that have been ruined by horses. Not fabulous horses, but nags. Tired, sad-looking nags, whose hooves go deeper than cows or sheep, and damage the land.

We should be using a lot more rough land like this to grow 'biomass'. Biomass is plant matter that provides a renewable source of energy and a storable, flexible fuel. Plant matter can be converted to heat and/or electricity using existing well-established technologies, like power stations, combined heat and power or heat alone – wood-burning stoves being a popular example. They are not just energy crops but also remove atmospheric carbon to store in the soil.

A favourite crop for biomass is *Miscanthus*, which has been developed, ready to use, as a low-carbon feedstock in the European bio-economy.[25] The EU sees biomass as an important component in their green energy mix.[26] *Miscanthus* differs from short-rotation coppice willow, an alternative energy crop, in that it gives an annual harvest and thus an annual income. At present it is only grown as far north as Yorkshire,[27] whereas willow coppice can be grown on what is now rough grazing land.

There have been schemes in the UK to encourage growing 250,000 acres of energy crops to help meet our targets to generate 15 per cent of our power with renewables by 2020 and the global warming targets.[28] Despite this, less land is used in the UK to produce biomass for burning than in 2009.[29] However, the government is encouraging biomass burning on a large scale. Most biomass burnt in the UK is wood, as the converted coal power stations, including Drax and in future Lynemouth, can only burn high-quality wood pellet. Around 15 million tonnes of wood were burnt in UK in 2015, of which only a fifth came from UK virgin wood, and a further 750,000m tonnes from waste wood. 'Drax burned imported pellets made from 9.1 million tonnes of wood. Most of those came from the southern US, with Canada and the Baltic States a joint second.'[30] Biomass energy generation also creates carbon emissions – when we should be saving them in the soil. There are some sustainability standards or 'land criteria' regarding biomass growing, but they are policed by the industry themselves, so shipping wood from the USA qualifies. According to *Biofuelwatch,* who keep an eye on the moves to big wood burning, even if the government did improve its standards, there might be a challenge in the WTO,[31] as it would be considered a 'barrier to international trade'. We must be able to produce more of our own trees.

Everybody is screaming that 'we need more houses, but not here please'. There is plenty of land that could be used to provide proper rural communities. Where I live, the council have just fined the latest housing developers for lack of drainage that led to some of the flooding on Boxing Day in 2015. We should do what the Confederation of Forest Industries suggest and include trees in all building permissions.

I have some audacious suggestions for change of land use that would be good for the land, and for us. Let's change a third of each of the following: moorland to forest, rough grazing to orchards, pasture to soft-fruit growth, pastures to fodder crops/rye, and arable to pasture – called mixed farming. We should also have more rotations – the best way of dealing with pest and disease build up.

But some farmers don't like rotation. The EU introduced the 'Three-Crop Rule', which stipulated that arable areas of more than 30ha must grow at least three crops or farmers would lose 30 per cent of their annual direct payment. One crop must not exceed 75 per cent of the arable area and two other crops should not be less than 5 per cent each.[32] These rules are aimed at benefitting the environment by encouraging biodiversity and rotation and reducing reliance on monocultures (see Chapter 7). Rotations are the best way to control pests – and have been used for centuries. But the former Environment Secretary, Andrea Leadsom, known as a leading Brexiteer rather than an agricultural scientist, dismissed the three-crop rule as simply ridiculous and bureaucratic. This is one of the regulations she wants to eliminate.[33]

SAVE OUR ARABLE LAND!

I am very pessimistic about whether we will be able to save our arable land. This ploughed land in the East of England is where our soils are suffering most, losing over 2 million tonnes of soil a year according to DEFRA.[34]

Erosion

Soil erosion is a major concern, as the then Chair of the Adaptation Subcommittee on Climate Change (CCC), Lord Krebs, spelt out in his introduction to their progress report. Soil erosion is a major threat to food supply and could mean that large areas in the East of England might become unprofitable within a generation.[35]

The committee's evaluation of the UK National Adaptation Programme, which aims to protect the countryside from a changing climate,[36] concludes that urgent action is needed. The rate of erosion has been the same for some time – 1–3 cm/year in East England – but the reasons and the evidence for erosion needs further research, according to Cranfield University.[37] The Natural Environment White Paper, the *Natural Choice*,[38] which updated *Safeguarding Our Soils*,[39] identified 5 (out of 7) ecosystems where soil quality has declined since 1990 and estimates that soil erosion is costing the country £150–250m/yr.

Carbon

The Countryside Survey Executive (Summary and Chapter 2) did not find any carbon decrease in most soils of the ten habitats – like woodland, bracken, fen and pasture. However there were significant losses from arable soils – those growing crops and weeds.[40]

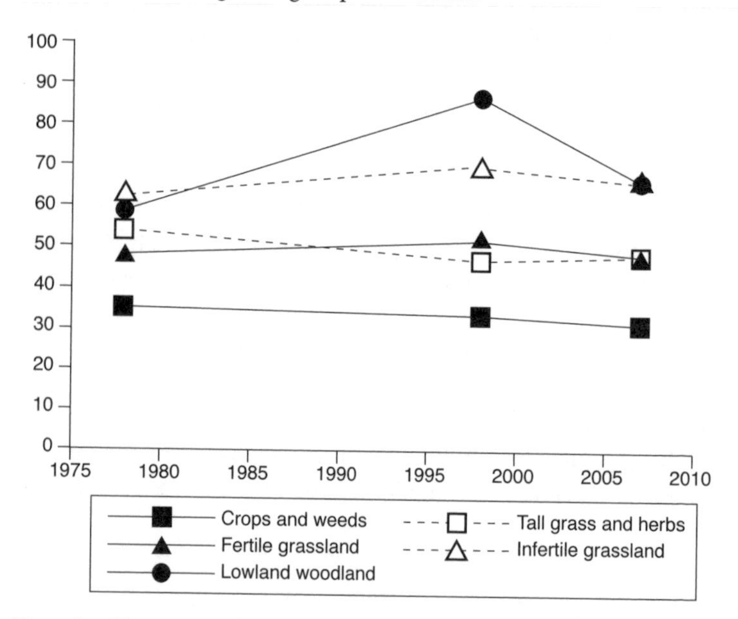

Figure 2 Change in Soil Carbon Content 1978, 1998, 2007 (Countryside Survey © Database Right/Copyright NERC– Centre for Ecology & Hydrology. All rights reserved. www.countrysidesurvey.org.uk/content/citing-use-countryside-survey-data-and-publications)

Note: 'Mean carbon concentration' is the average amount of carbon in the top 15cm of soil. That amount increased over the twenty odd years in both fertile grassland and lowland woodland, but decreased slightly under arable crops. See this chapter for more explanation.

The broad habitat called arable land shows statistically significant decreases in carbon over the last 30 years. The survey did not think global warming was the cause – although it could be an effect. The executive expressed concern that this was due to management practices. I cannot find any follow-up to this conclusion. Our cropping land has lost 17 per cent of its carbon in the last 30 years. That sounds serious to me. So I did some calculations, and if this loss translated to carbon dioxide emissions, it would represent a contribution of about 1–2 per cent of total UK Global Warming Potential (GWP).

I used various figures, some metric, some imperial, to try to assess carbon loss. The Countryside Survey (CS) figures are in carbon grams per kilogram soil, across the total 15.3m arable acres in UK.[41] These calculations are based on the weight of 8cm of topsoil (furrow depth) of 2,000,000lbs = 980kgm.[42] Using the CS table, these losses account for 7448 x 15m acres, i.e. c. 112,000mkg (or 112m metric tonnes) of carbon lost from the soil in last 30 years. If that carbon is translated into carbon dioxide, this amounts to 410m tonnes of CO_2 lost in the last 30 years from arable land in the UK, which equals 13.5m tonne/yr contribution. Current total UK e-CO_2 is 530mt/yr[43] so this is about 2.5 per cent of total greenhouse gas emissions. This is a significant amount. We should be taking it into our national calculations.

Heath Malcolm at the Centre for Ecology and Hydrology says: 'Greenhouse gas removals by soil from the atmosphere amounted to 15.5 MtCO2e in 2013, meaning that the net emissions from soil were 6.75 MtCO2e – 1.45% of the UK's total emissions.'[44] This would correspond to my calculations, as most carbon loss will be from arable land.

We hear of all sorts of greening technologies, especially in the energy sector reductions, but little about carbon loss from arable soil. I am willing for people to check my figures – as I don't think it should be me doing them anyway. I would like to see what DEFRA calculate, but there is a deafening silence. Yet if my figures are correct they should be declared to the International Panel on Climate Change (IPCC).

The Committee on Climate Change's Report to parliament in June 2015, 'Meeting Our Carbon Budgets Progress Report 2015',[45] mentions methane and nitrous oxides as contributing to global warming, but it doesn't mention carbon loss and possible carbon dioxide increase. In fact in the Headline Indicator table for emissions of agriculture, they say CO_2 is 'not applicable'.[46] The committee notes that to 'Strengthen the current voluntary approach to reduce agricultural emissions the farming industry should develop robust indicators to properly evaluate the Greenhouse Gases (GHG) Action Plan.'

Where Has All the Carbon Gone?

Most commentators' immediate reaction is to blame ploughing. Certainly, whenever you plough, carbon dioxide is released.[47] That is why 'no-tillage' systems are popular in organic systems. A law called the Environmental Impact Assessment (Agriculture)(England)(2) Regulations 2006 was introduced to control the ploughing up of pastures.[48] But has there been a concomitant increase in ploughing that could explain this carbon loss – or perhaps deeper ploughing? I asked some farmworker friends who are out in these fields all day:

Steve: 'with minimum tillage and strip-till, the decrease in ploughing has been immense, many farmers have sold their ploughs, those that do still plough only do so rotationally or behind grass. With more blackgrass (weed) coming, some may move back to ploughing.'

Matthew: 'ploughing has probably become less frequent in recent years. We run a plough-based system for both cereals and forage maize establishment, but many others locally are using either non-inversion or direct drilling. Non-inversion involves surface cultivation often combined with deeper subsoil loosening. Typically at least one pass to create a stale seed bed which is then sprayed to kill volunteers/weeds before drilling.'

Direct drilling typically involves no surface cultivation and seed is planted straight into the previous year's stubble. From the technical press, it is clear that ploughing is seen as both slower and more expensive by many farmers. The trend towards larger and larger arable units, together with a predominance of winter-established crops, has reduced the frequency of ploughing, although grass weed pressure and spray resistance has meant that some are returning to ploughing as part of a rotation. All of which seems to the point the finger away from ploughing.

The second option, and it is only a guess as nobody seems to be doing the research, is that with bigger yields, more food is taken off the land and less is being left behind to return carbon back to the soil. Somebody should do the measurements – other than Adam in the Archers. He has noticed that David Archer's fields have more worms! Hopefully soon, rather than just talking about worms, he could send some clods to the local university college – Harper Adams – where they could also look at the other soil creatures.

And that is where I come in. Forests house twice as many small soil creatures as pastures, which in turn house twice as many as arable land. While I was doing my PhD, looking at soil creatures, I found that herbicides kill off soil animals. It is obvious why. Herbicides do what they say on the tin – kill weeds. Each weed killed means there is less biomass for soil animals to live and feed on, and that means less soil life – and hence less carbon in the soil. We will look later at whether this may explain such a reduction of carbon in arable soils in Chapter 9.

SOIL HEALTH

The Environmental Audit Committee said the government needed a national monitoring scheme for soil health to ensure that there is adequate information available on the state of the nation's soil.[49] The government claimed that ministers were already taking action to improve soil health, but acknowledged that more needs to be done. It said a forthcoming 25-year Environmental Plan 'should set out specific, measurable and time-limited actions' towards increasing soil organic matter. That 25-year plan was shelved shortly after the Brexit vote.[50]

Soil Quality Indicators are properties of soil or plants that can be measured in order to provide clues about how well the soil is functioning. Indicators can be physical, chemical and biological properties, processes or characteristics of soils. They can also be morphological or visual features of plants.[51]

Chemical indicators include electrical conductivity, reactive carbon, soil nitrate, soil pH and extractable phosphorus and potassium. Sheffield University, comparing allotments and arable farms, 'measured a range of soil properties, including soil organic carbon levels, total nitrogen, and the ratio between carbon and nitrogen (which are all directly related to the amount and quality of organic matter in the soil) as well as soil bulk density, an indicator of soil compaction'.[52]

Physical indicators can measure bulk density, infiltration, soil structure and macropores, soil depth and water-holding capacity, as well as retention and transport of water and nutrients. Cranfield University carried out a study for DEFRA to find which Soil Quality Indicator[53] was best for determining quality of soil using physical indicators. Out of the top seven, they could not single out any particular indicator.

We should have indicators of the *living* soil – a way of shining a light on what is otherwise a dark and unseen world. We know more about planets billions of miles away than we do about what is under our feet. I'm not the first to say this. Leonardo de Vinci said: 'We know more about the movement of celestial bodies than about the soil underfoot.'[54]

Biological indicators include earthworms, microbial biomass C and N, particulate organic matter, soil enzymes, soil respiration and total organic, and small soil animals. In a Food Ethics Council Blog for Soil Day 2014, 'Save our Soils',[55] I said that 'soil organic matter' is probably a better measure than 'carbon content' when trying to assess the health of the soil. However, I think live animals are a better measure than dead carbon.

'Soil animals have an important role in the formation of soil structure. Soil animals improve soil structure by forming channels and pores, concentrating fine soil particles together into aggregates and by fragmenting and mixing organic matter through soil.'[56]

The idea of indicator species is well founded. In the UK the Biodiversity Action Plan (BAP)[57] sets out priority species that we need to keep an eye on: the so-called BAP species. We could have something similar to indicate the state of our soil health.

Soil Animals

This is the part of the book you won't find anywhere else – an introduction to the life of small soil animals and their role in the soil. I am one of the few remaining soil zoologists in the country, so feel responsible for revealing the last great animal kingdom to be explored.

According to the Countryside Survey, there are an estimated 12 quadrillion small soil animals in UK soils. In Brian-Cox-speak, that is 12,000 trillion. I counted half a million in three years. At that rate, it would take twice the age of the Earth to count the lot. They are mainly springtails, mites and nematodes.[58] Most soil animals are found in woodlands, where in any one soil sample, you will find several hundred arthropods consisting of 30–40 main species.[59]

Oribatids are small mites, yet possibly the most important group of animals on the planet – according to socio-biologist E.O. Wilson, who is one of most significant science commentators. He said in a *New Scientist* interview:[60]

> Saving Earth's biodiversity will take nothing less than an IPCC for species ... We are not making the headway we should be in preventing the destruction of ecosystems and species ... Most Americans have only the vaguest notion about any of that, even though they can talk intelligently about climate change. Yet when it comes to the living world they are in danger of losing something they scarcely understand. What are they missing? People see nature as trees, plants and vertebrates. Yet the world is run by little creatures most people have not heard of; 99 per cent of Earth's organisms are extremely small. For example, some of the most abundant and crucial land animals are the tiny oribatid mites, which are the size of a pinhead and look like a cross between a turtle and a spider. They are a linchpin organism of the environment, but 20 years ago when I set out to identify them no one had heard of them. Back then there were just two people in the US able to identify them. Fortunately one agreed to work with me. Yet we still don't know what the vast majority of oribatids do.

When I heard this I realised I should do more to publicise what goes on in the soil as I was one of the few people who used to be able to identify oribatids – although not anymore!

Table 4 My Suggestions for Soil Animal Indicators of Good Soil Health

Indicator	*Why*	*How*	*Role*	*Supporting Evidence*
Starling	V. visible	Eat worms	Provide food for higher up chain	Birds'-eye view[61]
Worm	Iconic status	Make soil structure	Provide spaces for other animals	Worms at work[62]
Springtails	Easy to extract	Look after roots	Keep plants healthy	Abundant[63]
Oribatid mites	E. O. Wilson	Break down debris	Transform inedible stuff into edible for others	Mites feeding[64]
Mesostigmatid mites	Like Buzzards	Eat springtails	Must be lots of springtails if these mites around	Mesostigmatid[65]
Woodlice	Easy to recognise	Eat fungal cords	By eating microbes keep GHGs down	Global warming[66]

The role of soil animals in global warming is little understood. Although the IPCC factor land into their calculations, they are predominantly physical and chemical factors. There is evidence that soil animals limit the amount of decomposition gases that would contribute to global warming by consuming microorganisms and thereby holding carbon in the soil – in living creatures.[67]

BIRTH OF THE EARTH

Soil animals were vital in the creation of soil. While this is something we can be sure Brexit will have no effect upon, I thought some of you might be interested, as it is quite an important and original theory.

Many people, including many evolutionary biologists, think that somehow plants and animals plopped out of the sea on to land at some point in geological history. But if they did, they would have landed on rocks. Reptiles can do that, but return to the water. Plant spores can do that to an extent. But there were no plant seeds that could do that until the plants colonised the land – when it was earth, not rocks.

If you consider the soil as a living entity, surely it must have evolved at some point. If you do believe it is alive, can you answer the next question: 'When did the soil evolve?' It wasn't always there. It wasn't there when our Earth came spinning off the sun 4,500 million years ago. There were just rocks and water, and later sludge. I reckon 90 per cent of the history of our Earth, was without earth.

My theory of soil evolution is that it happened around 350 million years ago, plus or minus 10 million years. I base this on the evolutionary age of all those small soil animals that I studied. All the time I was counting them, I knew that they were primitive creatures, as they pre-dated insects. Many are springtails, which used to be classed as insects but are now seen as the first hexapods – six-legged creatures. The other dominant arthropods in the soil – the mites – are pretty primitive creatures too. Insects came later. These creatures, along with worms and nematodes, created the first soils. Some – springtails – kept plants healthy, while others – the mites – dealt with plant remains.

I believe that the springtails came first from the sea, and kept the roots clean of debris and nasty fungi, while also encouraging good fungi for the plants. These fungi are called *Mycorrhiza* and increase transfer of nutrients into the plants, in exchange for energy from the plants. The springtails accidentally bring their fungal spores to the roots, acting like the bees of the soil – doing something useful accidentally. The soil mites probably evolved a bit later, as plant debris was washed to the oceans, and they came in to break that debris down.

Plants started growing in volcanic ash around the end of the Devonian period when two great continents collided to make one continent – Pangea. This period is known for its loss of fish and the evolution of fish with limbs, derived from fins, to move onto land. This terrestrial invasion was accompanied by the growth of 'higher', vascular plants, needing all the different minerals for roots stems, leaves and ultimately seeds. Up till then it was just ferns and mosses. Vascular plants can use water from the roots to keep the stem upright and the leaves open. This seems to have occurred because of the presence of volcanic ash. They need a range of essential elements for various parts of these processes, so need to be somewhere where *all* those elements are present. Volcanic ash has all those elements. The springtails came in from the sea, and kept the plant roots clean and introduced the spores to the roots. The fossil evidence shows that volcanic ash and plant debris were washed into the seas at that time, indicating that nothing was feeding on the plant debris. The oribatids arrived later to break down this debris, and recycle the nutrients that were otherwise washed to the sea. So the soil, higher plants and soil animals co-evolved.[68]

WHO SHOULD BE DOING WHAT?

This shows the complexity of the earth beneath our feet, providing the fundamental building blocks of the natural world. Yet our politicians are treating it – literally – like dirt. The Countryside Survey said in 2007 that soil management practices were poor, but nobody seems bothered. When the EU tried to introduce a Soil Framework Directive that required member states to come up with 'Action Plans', the NFU helped to block it and the DEFRA minister in the Labour government, Hilary Benn, said: 'The UK stated that it still had significant concerns regarding the cost and complexity of implementing the proposed directive and called for a different approach which recognised the differences between member states.' This attitude could cost us the earth.

The UK's role in the blocking minority prevented a qualified majority implementing the directive in EU, and it was withdrawn in 2014.[69] Soils are included in the Seventh Environment Action Programme of the EU. This provides that by 2020 'land is managed sustainably in the Union, soil is adequately protected' and leads to 'increasing efforts to reduce soil erosion and increase organic matter ... and to enhance the integration of land use aspects into coordinated decision-making, supported by the adoption of targets on soil and on land as a resource, and land planning objectives'. As a consequence of Brexit, it would be pleasantly ironic if the other 27 EU countries now re-introduce a new form of the directive. There is a campaign to do so.[70]

The UK government signed up to the '4 in a 1000'[71] campaign after the Paris Conference on Climate Change in 2016. This pledges to increase soil carbon by 0.4 per cent a year; this would mitigate all other GHG rises. I have yet to see any plans to develop that. 2015 was the International Year of the Soil, but you wouldn't have known that from DEFRA notices.[72]

The Environmental Audit Committee produced a report[73] on the state of UK soil health in 2016, before the Brexit vote, saying: 'the UK lacks an ongoing national-scale monitoring scheme for soil health ... successive

Table 5 EAC Recommendations and the Government's Response[74]

Recommendation 6 *'COP 21 signed up to increase soil carbon levels by 0.4 per cent a year for next 25 years. As part of the 25-year environment plan, it should set out specific, measurable and time-limited actions that will be taken to achieve this goal.'*	**Government Response** The government responded that it will continue to investigate ways to increase soil carbon and this will be part of the big picture of what we want our environment to look like post Brexit. It says there are limited opportunities under UK conditions to sequester carbon. They look at peat as main option, when they should be looking at arable land and moorland.
Recommendations 8 and 9 *'There is reason to doubt that the current cross compliance regime is achieving its goal of preventing soil damage. The Government should produce and consult on proposals to increase the ambition, scope and effectiveness of cross compliance.'*	**Government Response** The government stands by its belief that our soils are being protected through Environmental and Countryside Stewardship schemes saying: 'Good Agricultural and Environmental Conditions (GAECs) form part of the requirements under Cross Compliance and apply to anyone who receives payments under Single Payment Schemes and certain Rural Development schemes. These requirements apply in addition to underlying obligations under European and UK legislation.[75]

Recommendation 10 *'Maize production can damage soil health when managed incorrectly, and incentives for anaerobic digestion should be structured to reflect this. The double subsidy for maize produced for anaerobic digestion is counterproductive and has contributed to the increase in land used for maize production.'*	**My Response** Maize is perhaps the most soil-damaging arable crop – it is harvested late when the wet conditions for compaction are high.[76] Yet their only control is from the Red Tractor scheme that advises planting maize crops on flat land.[77]
Recommendation 12 *'The lack of monitoring prevents us from having nation-wide knowledge about trends in the health of our soil. This gap is not new, and successive Governments have ducked the challenge since the Royal Commission on Environmental Pollution recommended a national monitoring scheme in 1996.'*	**Government Response** 'Most soil properties change very slowly over time, so frequent monitoring is not justified and equally there is a substantial cost implication attached to monitoring.' The then government set out the aim in 2011 of managing soils sustainably and tackling degradation threats by 2030. We stand by that aim.
Farming Regulation Task Force February 2012 DEFRA (7.46): 'We recommend that DEFRA changes GAEC 1 to "a duty of care" to protect the soil and to prevent damaging soil erosion, reduce compaction, damage to landscape features and, over the long run, maintain organic matter in mineral soils'.	The government did not say 'Yes' – as they did for 159 of the other 200+ recommendations – but said it is 'Under consideration'. 'We will work with industry to explore the full range of options for the Soil Protection Review (SPR10) ... To start the process, we will launch a farmer survey in February 2012 to evaluate the implementation of the SPR10 so far and explore current soil management trends in England.'

Governments have neglected to establish a rolling scheme to monitor soil health. We heard that such a scheme could be affordable and would not be overly difficult to establish. We call on the Government to set up such a scheme and to explore whether innovations from Wales, involving alignment and co-funding with EU payments, could be rolled out to the rest of the country.' Obviously 'co-funding' is now out of the window!

Natural England 2015 says in its 'Summary of Evidence – Soils': '[soils] enable us to identify areas where the evidence is absent or complex, conflicting and/or contested. These summaries are for both internal and external use and will be regularly updated as new evidence emerges and more detailed reviews are complete.'[78] I thought the Countryside Survey had done that ten years previously, but nobody is taking any notice.

The 'Soil Survey of England and Wales' was established in 1939 with the aim of mapping, describing and classifying soils of the two countries. Re-named 'Soil Survey' and 'Land Research Centre' and then the National Soils Inventory (NSI) it is now part of Cranfield University.[79]

To summarise: we saw earlier that cross compliance/agri-environmental schemes/countryside stewardship, whatever you call them, are not effective, yet the government believes in them. The funding for them is part of CAP, so will go after Brexit. There are no directives, regulations, policies, plans or monitoring to protect the long-term health of soils under arable crops, and no sign that the government or the farmers' union want anything different.

LAND RESEARCH

'History ... celebrates the battle-fields whereon we meet our deaths, but scorns to speak of ploughed fields whereby we thrive; it knows the names of the king's bastards, but cannot tell the origin of wheat. That is the way of human folly.' This is a quote from *The Wonders of Instinct*,[80] written over 100 years ago by the fabulous and largely forgotten naturalist J. H. Fabre – known as the Insect Man.

Natural England collects a wide range of evidence to help manage soils, biodiversity, geodiversity and land management.[81] DEFRA have LandIS – Land information Service.[82] The Natural Environment Research Council (NERC) and the Biotechnical and Biological Sciences Research Council (BBSRC) are reviewing DEFRA's evidence base and the wider scientific literature on soil protection issues, to identify where the key risks lie. In 2017, spending included £1m on soil research in DEFRA's Sustainable Land and Soils programme, £5m investment by BBSRC on 'Soil and rhizosphere interactions for sustainable agri-ecosystems', £5m investment by the Natural Environment Research Council on Soil Sustainability and £2.3m on a Centre for Doctoral Training in Soil Science through BBSRC and NERC. ADHB are also funding research into soil health. The Technology Strategy Board, together with DEFRA, the BBSRC and the Scottish government, are investing up to £8.75m in helping businesses develop innovative measurement technologies for efficient agri-food systems. And the government announced in July 2013 it was allocating £160m[83] aimed at 'bringing science and agriculture closer together that could lead to a step-change in efficiency, profitability and resilience of our domestic farm businesses', through N8 AgriFood.[84]

I applaud this change in direction. The trouble is that many of the research stations that would have been asking the sorts of questions we need to address have been closed down.

Lost Land Research

I didn't realise how many research stations had gone until working as a Specialist Adviser to the Environment, Food and Rural Affairs Select Committee in 2008[85] on *Securing of Food Supplies to 2050*.[86] When asked which research station the MPs could visit, I thought 'Weed Research' in Oxford would be handy, only to find it had been closed down. How about Long Ashton, near Bristol? Closed. So I started counting …

It all started with a famous but elusive report – the 'Barnes Report'. Barnes was a MAFF man (Ministry of Agriculture, Fisheries and Food, that became DEFRA), who Mrs Thatcher sent round to research stations to draw a line between 'pure' research, that would be publicly funded, and 'practical/applied' research, to be funded by industry. As agriculture and horticulture are applied sciences, there wasn't much hope for the land-based research stations shown in Table 6.

Table 6 Land-based Research Stations Closed Since 1990

Animal Breeding RO	Hop Research Station (RS) at Wye	National Institute for Research in Dairying
Animal Virus RI	Houghton Poultry RS	National Institute of Agricultural Engineering Silsoe
Animal Disease RI	Institute for Research on Animal Diseases	National Vegetable Research Station (RS)
Food RI	Institute of Animal Physiology	Plant Breeding Institute
Glasshouse Crops RI	Letcombe Laboratory	Poultry RI
Grassland RI	Meat RI	Unit of Animal Genetics
Hannah RI	Monkswood Terrestrial Ecology	Unit of Nitrogen Fixation
Hill Farming RO	Macauley Institute for Soil Research	Welsh Plant Breeding Institute

Out of 32 research stations in the late 1980s – when we produced the most food ourselves – three-quarters of them have gone.[87]

Kenneth Mellanby, Director of the Terrestrial Ecology Research Station at Monkswood, known for his book *Can Britain Feed Itself?*, was my external examiner. Monkswood research gained an international reputation, in particular the work on pesticide poisoning effects on wildlife, on the value of rapidly disappearing hedgerows, and the establishment of a national Biological Records Centre. It was closed in 2009.

The Plant Breeding Institute (PBI) held a dominant position in the history of the UK wheat research/breeding industry for 75 years and helped Great Britain play a major role in global wheat research. It was sold to Unilever in the late 1980s, who sold it on to Monsanto in the 1990s. The National Institute of Agricultural Botany (NIAB) was also sold to Unilever.[88] They held all the patents and breeders' rights gained over 50 years of government research. Every seed lot sold today is inspected and certified by the NIAB, an independent, not-for-profit plant research station. It cannot receive public funds.

The Horticultural Research International Station at Wellesbourne (where the farm workers' union had been formed in the 1800s) was shut down almost entirely in 2010.[89] The Glasshouse Research Station at Littlehampton was closed and moved to Horticulture Research International (HRI) at Wellesbourne. HRI was government funded until the EU introduced the Single Farm Payment. The Labour government said they didn't need to fund it anymore, as horticulture would now get a thin slice of subsidy, so it was taken over by Warwick University in 2004. They closed it down five years later.[90] Prospect Union described it as 'Scientific Vandalism'.[91] The Warwick Crop Centre remains.[92] Other universities to kill off agricultural science include Imperial College, which shut down Wye College, and Plymouth University, which closed Seale-Hayne. Reading University do not now offer a degree in horticulture. There is only one place I think where you can get a horticultural degree – via Myerscough College in Lancashire. A lot rests on Harper Adams University. I asked one of its staff why they had been successful where others had not, and he said 'because we maintained independence'.

The Vegetable Seed Bank housed at Wellesbourne, originally funded with an Oxfam appeal[93] in the mid-1980s, was vulnerable at the time of the HRI closure. It was the vegetable bank for the UK, and for some vegetable (Allium) families, all of Europe. Curiously, at the same time,

Russia looked like doing away with their seed bank – the Vavilov Centre, the first ever seed bank. There, a dozen scientists starved to death in the Second World War to save their seeds getting into the hands of the Nazis. They knew how valuable this stuff is.

There are good people working in the remnants of these organisations, exemplified by people from Silsoe merging with Cranfield University. Currently they are training farmers and advisers in one-week short courses on the principles of soil and water management in order to meet the requirements of the European Union for sustainable soil and water management. I wonder what will happen to that, post Brexit. Their skills would be extremely useful in promoting better soil management.

Long Ashton Fruit Research developed home-grown sources of vitamin C during the Second World War, creating the brand Ribena. Weed Research and Letcombe Laboratory were merged into Long Ashton, but it was then closed in 2003. However, the overall truth is that a great diversity of land research has gone, and we have lost the chance to be leaders in this field (literally!), as we once were.

We should be making money from public sector discoveries. Imagine if there was a small royalty on Ribena, what research that could pay for! We should license public sector discoveries – not quite as tightly as private companies, but enough to pay for further research, something to plough back into land research. Take the famous 'Malling Rootstocks' that will be under all of your fruit trees, here and across the world. They were developed at East Malling Fruit Research station in Kent. For every £1 spent at Malling, going back to the First World War, Brookdale Consultants calculated £7.5 was returned to the economy. They reckoned that East Malling had provided over £9bn-worth of rootstocks globally. However, it never made any money from them. Instead, East Malling Research Station was 'rescued' in 2016, by … NIAB.[94] I recognise this is a simplification of complex issues around patenting of biological materials, but we should be having a debate about it.

While each closure is a loss, the greater loss is the whole research community that linked universities, colleges, research stations, ARC units and extension services directly to farmers. We exchanged information freely, to identify the problems and deal with them. The disappearance of the Agricultural Development Advisory Service as a free extension service is one of the great losses. Most investigations into the state of our present farming call for better bridges between research and farmers. There is a huge gulf between research and practice. In the last 20 years, land-based sciences, the field trials, have lost out to life sciences – carried out in the lab. We will see what that means in Chapter 9.

We are missing out on a whole range of skills. The Royal Agricultural Society of England said a few years ago that we do not have enough soil scientists to keep our soil healthy. When ash dieback hit, the Forestry Commission head of research said we did not have enough plant pathologists to deal with possible future invasions. It is one of the aspects of Brexit that we could, and should, introduce – stronger border controls on plants and animals to prevent introduction of new diseases. I once found yellow and black striped Colorado beetles – whose pictures used to be plastered in every police station and post office in the country asking us to watch out for them. They were on salad leaves in Berwick Street market in Soho, but when the MAFF man came he told me: 'don't worry they can't walk to potato fields in Hertfordshire from here'. We will have to be a whole lot more careful if we start importing more foodstuffs from different countries.

AHDB said in its submission to the Agri-tech consultation that 'there are a significant number of disciplines within the scope of Agri-tech that are now hardly taught at graduate level and where it will require positive intervention to build back national capacity. These include crop physiology, agronomy, agricultural engineering, soil science disease epidemiology, plant pathology, nematology, weed science, agricultural entomology, production horticulture, animal nutrition.'[95] The Royal Society, in *Reaping the Benefits,* added 'Universities should work with funding bodies to reverse the decline in subjects relevant to a sustainable intensification of food crop production, such as agronomy, plant physiology, pathology and general botany, soil science, environmental microbiology, weed science and entomology.'[96]

But the prospects look worse rather than better. Bio-scientists may become an endangered species.[97] They stand to lose £1bn/yr of funds from the EU, who say that once we have left they won't be funding bio-science in UK universities. These scientists may still be consulted, but will have less influence over EU legislation that applies to food and farm matters.

Questions

There are so many important questions that need to be addressed. The most obvious question is: 'how much carbon is being lost from UK arable lands?' Can somebody check my calculations – somebody from government should have done this already? Where has the lost carbon gone? Up in the air, out on trucks, or due to lack of weeds? We should not take a 'carbon reductionist' approach – i.e. just about the element called carbon. We should look at soil life, how soil ecosystems are degraded, which soil animals are most affected. Could we mitigate the effects of arable growing, by growing cover crop? Which of the arable crops protect soil life the best?

The big one I would like to see more research on is on the holding capacity of carbon under pasture compared with arable – i.e. animal versus vegetable? This is often left out when blaming animals for all farming's environmental ills. We should compare the soil degradation of intensively-farmed animals with pasture-fed ones. If 20 per cent of the maize crop now goes to animal feed, has this been factored in to comparisons with pasture-only rearing? That would include the health of soils under manured pasture compared with slurry-treated pasture. We need better statistics across a range of indicators. We need the figures not just the arguments. How much better is mixed farming for our soils? Could these changes in practice contribute to the promise for '4 in 1000' – where we try to capture 0.4 per cent extra carbon per year in the soil? Where pasture and arable alternate, how much better does it maintain soil carbon and soil life than monoculture arable soils? Does this make a better contribution to reducing GHGs than arable going organic? What about rotations? Can we demonstrate how much better rotations are for the soil? Is maize much worse for soils than – say – wheat and barley crops?

Another practice, popular at the moment, is 'mob grazing'. This is where higher concentrations of cattle/sheep graze pasture for shorter periods of time before being moved on. Some claim it is the answer, but again, we could do with good research. Does this make any meaningful contribution to carbon life and hence sequestration? Similarly, what are the benefits to the soil of waist-high wheat growing? The one I would really like to know – you'll see why in Chapter 9 – is how much carbon is lost due to weed-killer use? Each weed (i.e. plant) is a small carbon-capture and storage unit. Is the loss of carbon from arable land due to increased use of weed-killers? This sort of research can only be carried out by the state, as the likes of Monsanto are not going to do it. There are the whole range of research topics thrown up by agro-ecology, encouraged by both the FAO and the EU, which we'll pick up on in Chapter 7.

PART III

Farm and Food Science

7
Sustainability

In this chapter we address the issue of sustainability of food and farm systems. This relates primarily to the environmental impacts of production and consumption, both here and overseas. The crucial questions are whether we can keep treating the planet like this and whether we have opportunities post-Brexit to improve our performance. However, it is not just the planet affected but ourselves too. So we need to look at the best diets for both the planet and people.

The EU takes food and farm sustainability seriously. For agriculture, their first priority is: 'Creating a sustainable agricultural development path which means improving the quality of life in rural areas, ensuring enough food for present and future generations and generating sufficient income for farmers', by

> the promotion of agricultural practices and technologies that are environmentally sustainable and increasing rural incomes, such as integrated pest management, soil and water conservation methods, agro-ecological approaches and agro-forestry;
>
> improvement of access to productive assets such as land and capital and measures to ensure better delivery of essential services;
>
> initiatives improving income and reducing vulnerability for producers through capacity building and a comprehensive value chain approach.[1]

I can't see the 'promotion of agricultural practices that are environmentally sustainable' being near the top of our agenda. There is a Centre for Agro-ecology, Water and Resilience, near Coventry,[2] the Lincoln Institute for Agri-Food Technology aims to support sustainability, while Harper Adams University is promoting more sustainable food chains, Aberystwyth University prides itself on sustainable food credentials, and John Moores University has an MSc in Sustainable Food and Natural Resources. The universities of Cambridge, Lancaster, Exeter, Plymouth (who closed Seale-Hayne Agricultural College),

Worcester, St Andrews and Brighton all have strategies to provide more sustainable food.

The pressure will be to increase 'productivity' and 'efficiency', food security, and the curiously named 'sustainable intensification'.[3] This is where more food is produced on the best land, leaving the poorer stuff to grow more naturally. That is what always happens with capital investment – the best land is farmed too intensively, while poorer land doesn't get enough. Here I look at what we mean by 'more sustainable food', concentrating on the environmental impacts.

The global cropland 'footprint' associated with the UK food and feed supply increased by nearly 25 per cent from 1986 to 2009. Greenhouse gas emissions (GHGE) associated with fertiliser and manure application remained relatively constant. However factor in land-use change and GHGE increased over 10 per cent.

Here is the most important aspect. The UK is currently importing nearly 50 per cent of its food *and* animal feed. Yet, 70 per cent of the associated cropland and 65 per cent of GHGE impacts are located abroad.[4] While we are importing half our food, we are using twice as much of other people's land to grow that food, and two-thirds of our global warming impacts are occurring abroad because of these imports.

Let's look at the environmental impacts in terms of what goes *in* – land and energy – what goes *on* – in food production on air, water and nature – and what comes *out* – as waste.

LAND

In the world, as a whole, food and farming contribute about a fifth of the potential for global warming, according to the influential *Stern Report*.[5] This identified the three main contributors to GHGE as methane, nitrate fertilisers and, the most significant of all, land-use changes. Changes in land use take in a complex mixture of factors.

Land-use changes, according to that report, are the second largest source of emissions, after the power sector.[6] Land-use changes include that from forest, which absorbs and stores carbon in trees and roots, through to pasture land, grazed by animals whose muck is passed back to the land. To give a rough idea of the scale of the change, there are about half as many soil animals under pasture compared to trees – as we saw in the last chapter. The UK used to be heavily forested, but much was cut down several centuries ago.

The second major land-use change is from pasture to arable land. Using the same simple measure, there are about half the number of soil animals under arable land compared to pasture. These changes apply as much when we ploughed up pasture land in the East of England in the 1990s to what is happening in North and South America where ranches are ploughed up to make way for maize and soya. In Britain we now lose 2 million tonnes of soil eroded each year from arable lands. We need to factor in these two changes of land use – from forest to pasture, and pasture to arable – when calculating loss of carbon from the trees and soil, and also the increased emissions arising from the changes. Most calculations of global warming potential from food production miss these out. We hear only about methane emissions.

One of the main land pollution problems is from slurry – that is where cow manure is mixed with water from cleaning out the cowsheds. As more and more animals are reared intensively – around a fifth in the UK now – the problem of getting rid of slurry increases. Basically what was good healthy cow manure, going back to the soil, has now become 'a problem'. There are regulations to control slurry, whereas we used to welcome cowpats.

ENERGY

A calorie is the same the world over, and it is the amount of energy needed to raise 1gm of water through 1 degree centigrade. It is the same amount of energy, whether in food or fuel. In pre-industrial times, 1 calorie of energy produced 10 calories of food.[7] Now, according to Michael Pollan, for every calorie of food (or food-like substances, as he calls them) that is produced in the United States, 10 calories of fossil fuel energy are needed.[8] However this varies wildly. It takes roughly 25 times more energy to produce a pound of beef compared to a pound of corn. Roughly – measured in terms per person portion – pork needs 400 calories, beer 300, beef 250, bread 150, olive oil and oranges round 25. However, a carrot calorie from South Africa can account for 55 calories, whereas winter carrots grown in glasshouses account for 500 calories.

The farm and food system is highly dependent on oil, in particular to make nitrogen fertilisers. The Haber-Bosch process pressurises nitrogen with hydrogen (usually in methane) gas to make ammonia.[9] Manufacture of nitrogen fertilisers uses about 1–2 per cent of the world's total annual energy consumption.

The food sector is a major consumer of energy. The amount of energy necessary to cultivate, process, pack and deliver food to European citizens' tables accounts for about 26 per cent of the EU's total energy consumption.[10] The energy embedded in food products includes direct energy uses, such as moving a tractor, heating an oven or powering a mixer, as well as indirect energy flows, such as the energy needed to produce fertilisers or to operate irrigation systems.

Agriculture, including crop cultivation and animal-rearing, accounts for nearly one-third of the total energy consumed in the food production chain. The second most important phase of the food life-cycle is industrial processing, which, along with logistics and packaging (i.e. 'beyond the farm gate'), is responsible for nearly half of the total energy use in the food system.

The 'end of life' phase, including final disposal of food waste, represents only slightly more than 5 per cent of total energy use in the EU food system. However, that 'food waste' represents a proportion of all the energy use throughout the food chain – wasted.

These figures are good estimates, although it is difficult to be precise. It depends on what you count, and the form of energy involved. Some energy is better than others – e.g. biomass on farm compared to oil from abroad. There is high reliance in the food sector on fossil fuels, which account for almost 79 per cent of the energy. The relatively low share of renewables (solar, wind etc.) is because food consumed in the EU is imported from regions outside the EU where renewable is generally a low share of energy use. The same applies to the UK.

AIR

GHG emissions are similar to energy, in terms of proportion of food and farm contribution to overall environmental impacts. Food and farming contribute about 20 per cent of all UK GHG emissions according to Food Climate Research Network.[11] The *Stern Report* in 2007 (Appendix 7g) quotes similar figures for the world, saying farming GHG emissions are about 20 per cent. The UN says the same: in developed countries over half of that 20 per cent comes from food production, whereas in developing countries the proportion from farming contributes more than half.

31 per cent of those emissions come from methane emissions, caused by ruminants like sheep and cattle. According to Stern, 38 per cent of

farming's major contribution to GHGs comes not from animals, but from nitrogen fertilisers – used predominantly to grow crops.

Researchers at the University of Sheffield calculated that ammonium nitrate fertiliser accounts for 43 per cent of the GHGE produced during the manufacture of a supermarket loaf. This dwarfs all other processes in the supply chain. 100m tonnes of these fertilisers are used globally each year.[12]

There are about 50 UK policies relating to GHGEs and the food/farm sector,[13] along with EU policies on agriculture and climate change.[14] Agriculture is not a discrete part of the Emissions Trading Scheme (ETS), which expects every other area of business to do their bit to reduce emissions. Farmers in the EU do not trade greenhouse gases under the Kyoto Agreement. Yet they may harvest a good profit from lowering emissions, or sequestrating carbon in the soil. There is a proposed new ETS that does bind members to commit to agricultural GHG targets and the Irish Prime Minister pushed hard for agriculture to be a distinct sector for target purposes.[15] The European Commission suggested that the treatment of agriculture needs further analysis and that land use, land-use change and the forestry sector should be included. The DEFRA webpage that explained our role in EU ETS has been archived.

WATER

'A typical meat-eating, milk-guzzling, westerner consumes as much as a hundred times their own weight in water every day.'[16] Water is 'embedded' in all of the food we buy, wherever it comes from.

Like the land, we are using a lot of other people's water to produce our food. It takes anywhere from 1,000 and 5,500 litres to produce the food we eat in a day in Britain. By contrast, UK daily water use in the home averages 153 litres.[17] So, we directly use about a butt and a half each day, while others are using 30 times that to get our food to us. It takes 1,000 cups (140l) of water to make the coffee for your cup and 50 cups for a spoonful of sugar. It takes 2,000–5,000 litres of water to grow the kilogramme of rice you buy.

Tony Allen created the term 'virtual water'[18] to describe this usage. Virtual water is that required to grow the foodstuff. It is the irrigation needed and the water pumped to keep the crop or animals alive, plus any processing. A main part of the success of the 'Green Revolution' was built on increased irrigation systems. The world today grows twice as much food as a generation ago, but takes 3 times as much water to do so. Using this water to grow food threatens limited water supplies, as a lot of that water comes from pumps that are sucking up groundwater, creating the conditions for possible catastrophe in the future. It is estimated 20 Niles' worth of water is used each year in Africa to grow EU food.[19]

Coffee uses river and lake water (called 'blue' water), and tea predominantly relies on rain (green water). Kenya is a net exporter of 750bn metric tonnes/year of 'virtual' green water.[20] Ethiopia is another country profoundly affected by the desperate East African drought. While 98 per cent of its rainwater goes on home-consumed crops, it still 'exports' over 1,300 million cubic metres of virtual green water – the amount of rainfall needed to grow the coffee and oilseeds for export. No wonder they have drought problems. Various agencies are working to improve the 'drop per crop', but it is clear that won't solve matters.

Quite simply, water makes the world go round. As time goes on, water is going to become more and more important in the way we live our lives.[21] By 2025, water scarcity may cut global food production by more than the current US grain harvest – over 50m tonnes each year.

NITROGEN

Nitrate fertilisers can be credited for feeding probably an extra billion people on earth. Yet the United Nations Environment Programme says the nitrogen cycle is in hyperdrive and is one of three key areas where 'planetary boundaries' have been crossed.[22] These boundaries are the 'red lines' we should not cross as they threaten Earth's habitability. The other two boundaries are global warming and biodiversity loss. The UN set a boundary level that we should not exceed for nitrogen. The world is 3 times above that level. An estimated 120m tonnes of atmospheric nitrogen per year are converted into reactive forms, mainly the manufacturing of fertilisers and the cultivation of leguminous crops such as soybeans.

Manufacturing and transport of fertiliser are estimated at 6.7 tonnes CO2e for each tonne of nitrogen fertiliser. We use 850,000 tonnes in the UK,[23] equivalent to 5.7mtonnes CO2e. Out of the total UK GHGs of about 180m tonnes,[24] this amounts to about 3 per cent of total GHG contribution. Much nitrogen fertiliser production is based abroad (East Europe), with about 800,000 tonnes imported into the UK.[25] We must add to this the nitrous emissions (NO_x) following fertiliser application.

In the UK, NO_x are 6 per cent of UK greenhouse gas emissions.[26] Agriculture contributes two-thirds of this – a further 4 per cent. This means 7 per cent of total UK GHGs are due to nitrate fertilisers. But we hear nothing about this.

The main water pollution concern arising from farming in Britain is nitrogen pollution of our waterways, from run-off from fields. This stimulates algae growth and can choke off oxygen essential for other creatures in the river. The clever word for this is 'eutrophication'. The EU created the Nitrates Directive to control this pollution. Each member state has to assess and control the potential for pollution of waters with nitrogenous compounds generated from agricultural sources, and assess all waters every four years. DEFRA research is looking to see how quickly systems recover from nitrate pollution to evaluate the effectiveness of existing measures. Nearly 60 per cent of England is covered by Nitrate Vulnerable Zones (NVZs).[27] The government says: 'In the longer term and depending on our exit agreement, leaving the EU may offer opportunities to tackle nitrate pollution in a different way without compromising environmental protection.'[28]

PESTICIDES

The total external cost of pesticide use in the UK is in the range of £430m per annum, according to the government.[29] Globally, pesticide production releases 72m tonnes of GHGs, which is about one-sixth of fertiliser GHGs. We will look in more detail at their impacts in Chapter 9.

BIODIVERSITY

The word 'biodiversity' gives an indication of the variety and diversity of organisms in all sorts of locations. A wide diversity encourages all sorts of living webs at all levels from cells to the planet. The main concern is that we are losing 'biodiversity' in most places, most of the time. The greater variety of organisms you have, then the better resilience there will be. If one lot don't resist a pest, then another will. If you have a limited variability, once succumbed there is no recovery. Agrobiodiversity is a subset of biodiversity covering selection processes and inventive developments as well as crops, biocontrol agents, wild species, cultural knowledge and soil organisms.[30] Most of the main crops now grown come from narrow gene pools. All South American coffee and all

Malaysian rubber derive from four plants – not the same ones, of course. The narrow diversity and vulnerability is now affecting bananas, derived from a few at Chatsworth House in Cheshire, as we saw in Chapter 3.

1 Crops and Creatures

The FAO estimates 75 per cent of crop diversity was lost between 1900 and 2000. The *State of the World's Plant Genetic Resources for Food and Agriculture* predicts that as much as 22 per cent of the wild relatives of important food crops such as peanuts, potatoes and beans will disappear by 2055 because of changing climate.

The FAO estimates that there are roughly a quarter of a million plant varieties available for agriculture, but less than 3 per cent of these are in use today. With disuse comes neglect and possibly extinction. Modern farming concentrates on a small number of varieties designed for intensive farming. This has dramatically reduced the diversity of plants available for research and development and is known as 'genetic erosion'. The world's food supply depends on about 150 plant species. Of these, just 12 provide three-quarters of the world's food. More than half of the world's food energy comes from a limited number of varieties of three 'mega-crops': rice, wheat and maize.

Of the 8,800 animal breeds that have been known to humanity, 7 per cent are already extinct and 17 per cent are at risk of extinction. Of the over 80,000 tree species, less than 1 per cent has been studied for potential use. Fish provide 20 per cent of animal protein to about 3 billion people, but only 10 species provide about 30 per cent of marine capture fisheries and 10 species provide about 50 per cent of aquaculture production. Over 80 per cent of the human diet is provided by plants. Only 5 cereal crops provide 60 per cent of energy intake.

2 Food Products

We have 2,200 varieties of apple in Britain. But try to find a Peasgood's, Port Wine Kernel or Lord Lambourne at your local supermarket and you will struggle.[31] Just two varieties – Gala (28 per cent) and Braeburn (19 per cent) – now account for almost half of all sales across UK outlets. The market remains dominated by cheap imports and is focused on a few easy-to-grow varieties with long shelf-lives that travel well.

Most people have never eaten a decent pear in their lives[32] yet there are 550 traditional varieties, such as Coppy, Ducksbarn and Yellow Huffcap. Common Ground[33] campaigns to encourage diversity in orchards. Over half of all our pear orchards have disappeared in last 30 years, while 80 per cent of pears are now imported.

3 Seed Banks

The first seed bank was set up by perhaps the greatest ever agricultural scientist, the Russian N.I. Vavilov. He was proud to be both a scientist and a socialist. Despite doing more than anybody in history to reduce starvation, he himself died of starvation in prison at the hands of Stalin. He was accused of being 'bourgeois' because he believed in the genetic approach to breeding. As a result, he was cast aside for Lysenko and his 'Lamarckian' ideas.

There are now some 1,750 gene banks worldwide, with over 100 of them holding more than 10,000 accessions. There is a back-up bank, the Svalbald Global Seed Vault, in Norway. Of the total 7.4m samples conserved worldwide, national government gene banks conserve about 6.6m. 45 per cent of these are held in only seven countries, down from twelve countries in 1996. This increasing concentration of preserved genetic diversity in fewer countries and research centres needs attention by the FAO's International Treaty for Plant Genetic Resources for Food and Agriculture.

At present, variety registration and seed-marketing occurs across the EU. Seed royalties on UK-grown crop varieties are now collected under an EU-wide system. Presumably the NIAB will be closely involved in the transfer of plant rights, and it will be interesting to see what role the seed bank at Wellesbourne will play, as a consequence.

4 Biodiversity Action Plan

The UK has a Biodiversity Action Plan (BAP),[34] which identifies various species as indicators of biodiversity and at risk. The Biodiversity *Framework* was developed in response to two main drivers: the UN's Convention on Biological Diversity's (CBD) *Strategic Plan for Biodiversity 2011–2020* and the EU Biodiversity Strategy (EUBS), 2011.

Each species has an action plan. There are bird BAP species like lapwings, skylarks and grey partridge; fish species like Atlantic salmon and sturgeon. But two species do not make biodiversity. I don't know of any BAP soil animals.

WASTE

Food waste or food loss is food that is discarded or uneaten. The causes of food waste or loss are numerous, and occur at the stages of production, processing, retailing and consumption.

A third of food is wasted globally according to the UN, which accounts for 1 in 4 calories, at a time of malnutrition and hunger. Love Food, Hate Waste[35] say we throw away 7m tonnes of food and drink every year, of which half could have been eaten. Wasting this food costs the average household £470 a year, rising to £700 for a family with children, the equivalent of around £60 a month.

The Waste Resources Action Programme (WRAP)[36] analysed where the main food waste occurs in the food supply chain, and considers this could be halved and would help the deliver *Courtauld Commitment*[37] by 2025,which aims to make UK food and drink production more sustainable. By weight, household food waste makes up 70 per cent of post-farm-gate total. In addition to food ending up as waste, 710,000 tonnes of food surplus from manufacturing and retail is either redistributed via charitable and commercial routes (47,000 tonnes in 2015), or diverted to produce animal feed (2.2m tonnes).

The great indictment of our food system is that this waste goes on while many people on low incomes or in poverty are unable to afford 'healthy' foods and face rising food prices. Food prices have risen since the 2016 Referendum and will continue to do so as long as we are importing nearly half our food and sterling is plummeting. We have 2,000 food banks handing out 1.2m food packs a year. This accounts for only a small part of the total number of people affected by household food insecurity in the UK, often called 'food poverty'. Several supermarkets in Lancashire now leave excess food in cupboards called 'larders' so that people can collect food for themselves without the humiliation of going to food banks. The term 'food poverty' is not accepted by DEFRA, probably because it conveys what it is all about in two words. An estimated 5.8 million people live in food poverty. The impact of such poverty on people's health and well-being though lack of access to healthy food is far ranging.[38]

There are many local schemes that address this contradiction. The Soil Association reckons 20–40 per cent of fruit and vegetables are rejected on cosmetic grounds before they reach the consumer. There is a Gleaning Network[39] where people go into the fields and collect the fruit and vegetables left there and take the fresh produce to where it is most needed. Our Sustainable Food Lancashire group is involved in setting up a Real Junk Food Café, which welcomes everyone through its doors irrespective of whether they can pay.[40]

France is leading the way in addressing this issue with a law introduced in 2016.[41] After a grassroots campaign, supermarkets face hefty fines if caught chucking away edible food or pouring bleach on leftovers to deter dumpster divers. Instead, they have to send surplus groceries to charities. EU member states are committed to meeting the UN's Sustainable Development Goals (SDG) 2015,[42] by having a target to halve per capita food waste at the retail and consumer level by 2030. The plan is to measure, monitor and work out methods to reduce food waste.[43]

Put in environmental terms, if food waste were a country, it would rank behind only the US and China for greenhouse gas emissions.[44] The production of wasted food also uses around 1.4bn hectares of land – 28 per cent of the world's agricultural area.[45] Much food waste goes to landfill, where it rots and ferments, attracting all those gulls. Many council food waste disposal systems, like ours in Pendle, closed in the first round of austerity measures.

There are government initiatives on a variety of these environmental impacts, but not much in the way of a coordinated programme for sustainable food. There are funds for food security, but that often just means 'more food'. There is nothing to support that EU statement on agro-ecology.

Sustainability is hard to sell to individuals. 'Carbon labelling' never caught on, as nobody understood it. The main response of concerned customers is to go vegetarian. However, that still depends on how the vegetables are grown. Local stuff grown in a cooperative is different from that grown in plantations, flown in from Brazil or Kenya. Many blame *meat* for our ills, but it is more to do with the *means* of production. It is not just *what* we produce, but *how* we produce it.

There are many deep-seated issues when it comes to our food. It is one thing to have more sustainable food; it is another to eat it …

SUSTAINABLE DIETS

We can have all the sustainable food in the world, but we then have to eat it. Not just the occasional meal, but interesting diets that people want to eat. Diets refer to patterns of eating. There seem to be all sorts of diets claiming to be healthy and environmentally friendly.

One study looked at 14 common sustainable dietary patterns, where there were reductions as high as 70–80 per cent of GHG emissions and 50 per cent of water use. Reductions in environmental footprints were generally proportional to the reduction of meat consumption. Dietary shifts also yielded modest benefits in all-cause mortality risk. Environmental and health benefits are possible by changing current Western diets to a variety of more sustainable dietary patterns.[46]

There is only one study looking at sustainable diets that takes into account the quality of land (see grades in Chapter 6) used to produce crops compared with that used by animals.[47] The study calculated human carrying capacity (i.e. how many people the country can carry) under 10 diet scenarios. The scenarios included 2 reference diets based on actual consumption, along with 8 'healthy diets' that complied with US dietary guidelines but varied in the level of meat content. Annual per capita land requirements ranged from 0.5 acres to 2 acres per person/year across the 10 diet scenarios. Carrying capacity of the US varied from 402 to 807 million persons, which is 1.3 to 2.6 times the US population. Carrying capacity was higher with less meat and highest for the lacto-vegetarian diet, but the carrying capacity of the vegan diet was lower than the 2 omnivore diets. That is because vegetable crops need the best land for growing.

If sustainable diets are to be properly sustainable, we must 'sustain' our population better in terms of our own health. We look at how we factor health into our sustainable diets in the next chapter.

8
Obesity

Whatever happens with Brexit, one thing is sure – we have to do something about obesity. In this chapter we will look at what is going wrong and see how to rectify the situation. This has to be included in any plan for the future of food and farming, and cannot be avoided any longer. We are at greater risk from diet-related diseases, particularly diabetes, than any other non-transmissible disorders. But some will find the conclusion hard to swallow.

Proportionally, we have more obese people here than any other country in Europe.[1]

The EU has an Action Plan to 2020, aimed at halting the rise in obesity among children up to 18 years old. There is also a strategy on nutrition dating back 10 years.[2] However, actions are very much down to national level.[3] Nevertheless, if we are going to create a better food system, we have to tackle obesity now.

Following Brexit, there are concerns from health campaigners that with the government focusing attention on Brexit for the next few years, health initiatives will get forgotten. In particular, Jennifer Rosborough, campaign manager for *Action on Sugar*, said: 'We urge the Prime Minister [then David Cameron] to do the right thing and announce details of his much awaited childhood obesity strategy before his exit and leave behind a lasting legacy for the future health of the nation.'[4] The strategy was announced in August 2016 but met with a disappointing reception. The King's Fund said it ignored the Health Select Committee's recommendations.[5] Even Sainsbury's and the British Retail Consortium criticised it for not setting compulsory and measurable targets.[6]

The projections are that by 2030 in the UK, there will be 11m more obese adults, costing a further £2bn in health-care costs.[7] Obesity is a serious public health problem, as it significantly increases the risk of chronic diseases such as cardiovascular disease, type 2 diabetes, hypertension, coronary heart disease and certain cancers. This may lead to a range of psychological problems. For society, the substantial direct and indirect costs put a considerable strain on health-care and social resources.

Obesity is the main cause of type 2 diabetes. It accounts for 80–85 per cent of the overall risk of developing type 2 diabetes, throughout the world. People with type 2 diabetes in England and Wales are a third more likely to die prematurely. It is currently estimated that about £10bn a year is spent by the NHS on diabetes – about 10 per cent of the total NHS budget, or around £192m a week.[8]

We have to curb the rise in obesity. It is also clear that the nutritional guidelines are not working. Some will say that this means they should be 'imposed' more rigorously, while others believe the guidelines themselves are wrong. This is a contentious issue, on which I will make my own position clear. Whatever the truth of the matter, there should be a much more open debate.

FATS OR CARBS?

You may be forgiven for thinking that obesity is the result of eating too much fat. 'Fat makes you fat' sounds pretty obvious. Everywhere you go in the supermarkets, there are 'low fat' yogurts and milks, 'healthy' alternatives to butter, even 'healthier' cornflakes and crisps. The dominant mantra over the last 20 years has been to eat less fat.

A study in 2014 presented in *Annals of Internal Medicine*,[9] led by Cambridge University, analysed data from 72 unique studies with over 600,000 participants from 18 nations. It showed that the 'evidence does not support guidelines restricting saturated fatty acid consumption to reduce coronary risk nor does it support high consumption of polyunsaturated fats – such as omega 3 or omega 6 – to reduce coronary heart disease.'

The *American Journal of Clinical Nutrition* (2010) says: '21 prospective epidemiologic studies showed that there is no significant evidence for concluding that dietary saturated fat is associated with an increased risk of CHD (Coronary Heart Disease) or CVD.'[10]

According to the National Obesity Forum we should 'Eat fat, cut the carbs and avoid snacking to reverse obesity and type 2 diabetes.'[11] Their report, issued jointly with the Public Health Collaboration, was criticised by many scientists.[12] I largely agree with it, so we'd better look at it in more detail.

An article in *New Scientist* which looked at the issue[13] details the way in which we get a 'sugar rush' whenever we eat carbohydrates. These 'carbs' break down to sugars (sucrose) quite quickly – slower if they are wholegrain. Once sugar is in the bloodstream, our bodies secrete the hormone 'insulin' to remove the sugar from the blood and store the energy as fat in cells. Consuming too much sugar for prolonged periods can render insulin less effective, leading to type 2 diabetes. There is nothing like such a metabolic path for fats, which are broken down into fatty acids in the gut and are not as easily absorbed. This leaves us feeling 'full' for longer.

Try it. For breakfast, eat a bacon butty or an omelette. If you do get peckish, drink a full milk cappuccino. It will keep you going. Eat a bowl of cornflakes and you are peckish an hour or so later. In the evening, instead of biscuits go for nuts – and instead of crisps, go for pork scratchings.

Better start digging. There may be an entire Mediterranean diet buried under here

Flawed or Fraud?

These 'low fat' guidelines appeared in the late 1970s, and arose from a report by Senator McGovern, who was a Democrat concerned at the amount of hunger amongst plenty in the US. However, after about 10 years he turned towards the nutritional aspects – despite complaints from nutritional quarters that this was not his territory.

In January 1977, the McGovern committee issued a set of nutritional guidelines for Americans. These sought to combat killer conditions such as heart disease, certain cancers, strokes, high blood pressure, obesity, diabetes and arteriosclerosis.

Central to his recommendations was a study carried out by Ancel Keys called the Seven Countries Study,[14] still quoted to this day.[15] Keys produced a powerful graph (see Figure 3). Take one look at the strong line going up with any increase in fat consumption (as per cent of energy intake) linked with heart disease.

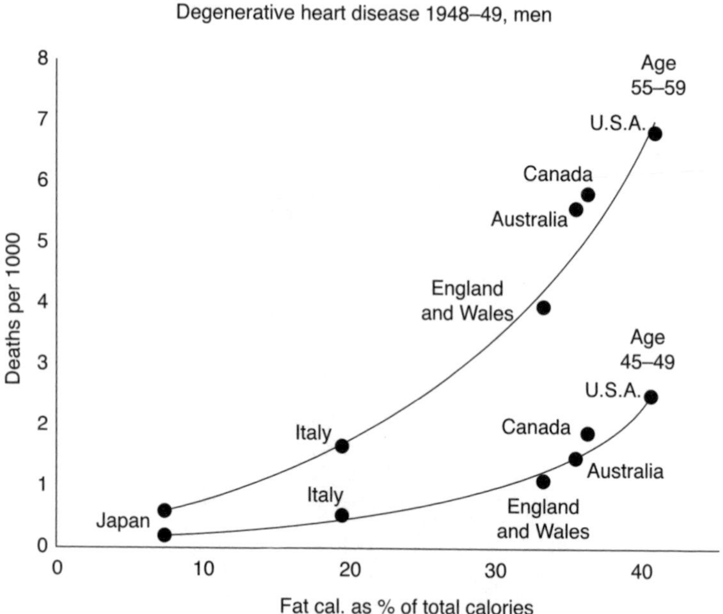

Figure 3 Ancel Keys 7 Countries Study. Note: this graph, produced by Ancel Keys, was influential in setting dietary guidelines.

It looks very clear – 10 per cent of calories taken as fat produces about 0.5 deaths from heart per 1,000 (Japan), whereas 40 per cent produces about 15 times that number (USA). That seems to fit – Americans are obese because they eat too much fat, whereas the Japanese, primarily fish-eaters, do much better. Italy benefits from the Mediterranean diet. No wonder people accepted this finding.

But Keys left out countries where people eat a lot of fat but have little heart disease, such as Holland and Norway, and countries where fat consumption is low but the rate of heart disease is high, such as Chile. Yurushalmy and Hilleboe used similar data to produce a graph based on 22 countries,[16] that contradicted Keys' findings (see Figure 4). Was Ancel Keys' work 'flawed' or a 'fraud'?[17]

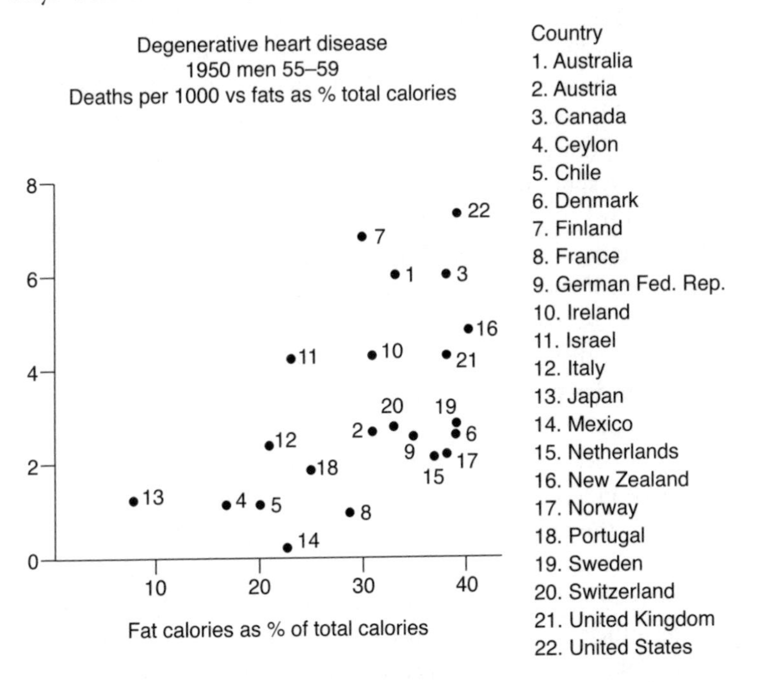

Figure 4 Yerashalmy & Hillieboe 22 Countries Study. Note: this graph – based on similar data to Ancel Keys but adding a further 15 countries – does not show such a clear line between deaths due to heart disease and fat consumption (% of total calorie intake). See Chapter 8 on obesity for more details.

Saccharine Disease

There was criticism of Keys' findings.[18] George Campbell and Thomas Cleave published *Diabetes, Coronary Thrombosis and the Saccharine Disease*, arguing that the chronic Western diseases such as diabetes, heart disease, obesity, peptic ulcers and appendicitis are caused by one thing – 'refined carbohydrate disease'.[19] Cleave called this the 'Saccharine disease'.[20] In 1957 the American Heart Association said: 'There is not enough evidence available to permit a rigid stand on what the relationship is between nutrition, particularly the fat content of the diet, and atherosclerosis and coronary heart disease.'[21]

The debate continued in the 1970s, although a Henry Blackburn said there was much talk from each side but little listening. John Yudkin retired to write *Pure, White and Deadly*.[22] He wrote:

> The first proponent of the idea that fat could be a cause of coronary thrombosis, and since then its most vigorous defender, was Dr Ancel Keys of Minneapolis. In 1953 he drew attention to the fact that there was a highly suggestive relationship between the intake of fat in six different countries and their death rates from coronary disease. It has been responsible for an avalanche of reports by other research workers throughout the world; it has changed the diets of hundreds of thousands of people; and it has made huge sums of money for producers of foods that are incorporated into these special diets. There is a sizeable minority of research workers, of whom I am one, who believe that coronary disease is not largely due to fat in the diet.
>
> It seemed appropriate to look much more closely at the figures of mortality and fat consumption … I found that there was a moderate but by no means excellent relationship between fat consumption and coronary mortality, which did not become closer even when one separated the fats into animal and vegetable. A better relationship turned out to exist between sugar consumption and coronary mortality in a variety of countries. The best relationship of all existed between the rise in the number of reported coronary deaths in the UK and the rise in the number of radio and television sets … If only a small fraction of what we know about the effects of sugar were to be revealed in relation to any other material used as a food additive, that material would promptly be banned.

According to Ian Leslie, prominent nutritionists combined with the food industry to discredit Yudkin's findings and destroy his reputation. His career never recovered and he died largely forgotten.[23] In recent years, however, Yudkin's reputation has been resurrected by Robert Lustig who has updated Yudkin's work.[24] In the meantime, people have reduced their intake of animal fat and cholesterol, and the incidence of obesity and serious disease has increased.[25]

DIETARY GUIDELINES

Like all science, it is not just a matter of molecules – there are other factors at work. We saw in Chapter 1 how the Republican Earl Butz increased the production of corn and soya to keep food prices low in the run-up to an election, but then had to dispose of the surplus by promoting cornflakes and turning corn into High Fructose Corn Syrup (HFCS). Fructose sugar is worse than common sugar, sucrose, because instead of ending up as fat in cells, it ends up in the liver. If one single person could be held responsible for obesity (which they can't) then Butz would be the one.

Opposing Butz was the Democrat Senator McGovern – who also wanted to be sweet with the corn farmers. He introduced the first 'dietary guidelines', based on Ancel Keys 'flawed' research. These dietary guidelines were 'manna from heaven' for the corn farmer, encouraging us to eat more corn products – rather than fat products produced by the ranchers. In 1976, just weeks before a tight presidential election, Butz left the USDA in disgrace after making a stunningly crude racist remark.[26]

When these dietary guidelines were introduced in the USA, it was not without argument. A number of scientists said in a Congressional hearing that the findings were not based on sufficient evidence.[27] They were ignored.

The US Department of Agriculture's (USDA's) dietary guidelines recommend that fats and oils be eaten 'sparingly'. The low-fat gospel spreads further by a kind of societal osmosis. Yet the UK diabetes organisation says the low-fat guidelines should never have been introduced.[28]

About 15 years after the US dietary guidelines were first published, the UK introduced similar guidelines. But according to the Open Heart Report of the *British Medical Journal*: 'There was no evidence for the introduction of the dietary guidelines, there was no need for them. Government dietary fat recommendations were untested in any trial prior to being introduced.'[29]

Since they were introduced,[30] the Body Mass Index (BMI) measure of fatness has gone up inexorably (see Figure 5). The latest version is the Eatwell Guide, which recommends consuming around 20gm of saturated fats for women and 30gm for men.[31] This amounts to about 10 per cent of the energy intake we need.[32] It is assumed that we will meet the remaining 90 per cent of our energy requirements by eating 'sensibly'. But most people believe 'low fat' is what is important, so they follow the low-fat signs to crisps, biscuits, cereals and other refined carbohydrates, and lots more sugar. A bowl of cereal can equal 9 spoonfuls of sugar, but parents think it is healthy.

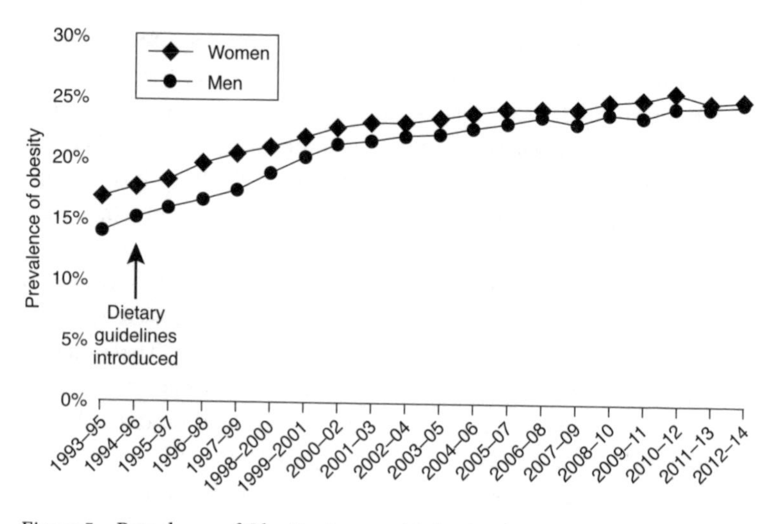

Figure 5 Prevalence of Obesity Among Adults Aged 16+ years. Health survey for Public Health England 1993–2014 (3yr averages) (http://webarchive.nationalarchives.gov.uk/20170110171021/https://www.noo.org.uk/NOO_about_obesity/adult_obesity/UK_prevalence_and_trends)

These guidelines are appearing all over the world – just as obesity is on the rise globally.[33] China approved a similar set of guidelines, using a 'Food Guide Pagoda', divided into five levels of recommended consumption.

1. Cereals (in the form of rice, corn, bread, noodles and crackers) and tubers make up the base of the pagoda and,
2. Vegetables and fruits (on the second level) make up the majority of any meal.
3. Meat, poultry, fish, shrimps, and eggs (on the third level) should be eaten regularly in small quantities.
4. A recommendation to eat milk, dairy products, beans and bean products is included in the fourth level.
5. Fat, oil and salt, placed on the roof of the pagoda, should be eaten in moderation. Recommendations to drink plenty of water and to do physical activity are also included.

This is a recipe for obesity and diabetes. China wants its workers to eat cheaply, and carbs supply that need. If they were all eating wholemeal, I'd be less concerned, as the fibre takes time to digest, but they still break down into sugars. Yet most carbs do not come like that, they come 'refined'. Look at the dominance of white sliced bread, biscuits, crisps and the like.

I see it when I walk round the supermarket. There are rows of shelves full of 'healthy cereals' and 'healthy crisps' that do you no good at all. No wonder people are confused and no wonder people who are making an effort to reduce weight often end up being frustrated.

Some call fat the 'greasy killer'. 'In America, we no longer fear God or the communists, but we fear fat', says David Kritchevs.[34] Gary Taubes wrote a series of articles in the *New York Times* called[35] 'A Big Fat Lie'. He said: 'During the past 30 years, the concept of eating healthy in America has become synonymous with avoiding dietary fat. The creation and marketing of reduced-fat food products has become big business; over 15,000 have appeared on supermarket shelves. Indeed, an entire research industry has arisen to create palatable non-fat fat substitutes, and the food industry now spends billions of dollars yearly selling the less-fat-is-good-health message.'

The latest dietary guidelines[36] for Americans update the older ones, but still see themselves as 'an essential resource for health professionals and policymakers as they design and implement food and nutrition programs that feed the American people'. They also provide information that helps Americans make healthy choices for themselves and their families. The 2015 Scientific Report on the guidelines[37] recognises that eggs are not bad for you – after all these years! It also notes that there is no evidence to support the old 'low-fat dietary guidelines'. Nevertheless, the latest US guidelines carry on as if nothing has happened, sticking to a maximum of 10 per cent intake of calories as fat.[38]

Meanwhile, the European Food Based Dietary Guidelines (FBDG)[39] are developed through the EFSA committee – the European Scientific Committee on Food Safety. They are a lot less prescriptive than the US and UK guidelines and contain straightforward information on healthy eating, aimed at the general public. They give an indication of what a person should be eating in terms of foods rather than nutrients, and provide a basic framework to use when planning meals or daily menus.

The WHO 2017 Sugar Guidelines suggest that we reduce our intake of 'free' sugars to below 10 per cent of total energy intake. This means to below 10 per cent of the total calories we consume each day: about 12 spoonfuls a day. Free sugars are those added to our drinks, biscuits, jams, syrups and fruit juices – not those in milk, whole fruit or vegetables.[40]

HARD TO SWALLOW

There is no single cause of obesity; it is a complex matter living in an 'obesogenic' environment. Yet I find it hard to work out why this fat v carb debate is only slowly coming into public view, and the way in which many nutritionists are dealing – or rather not dealing with it. Typical of the way the issue is being confused is Dr Phil Whitaker, writing in the *New Statesman*: 'Carbohydrate. Fat: it doesn't really matter. Eat less and do more. It's as simple – and as complicated – as that.'[41] No it is not. The idea that it is all about calories in and calories out is true only in terms of physics – not in biochemistry. Fats and sugars behave very differently. Sugar can enter the bloodstream very quickly, at virtually no energy cost to our bodies. But if we eat saturated fats, we burn calories to chew, break down the fat in the gut, and form fatty acids before they enter the bloodstream.[42] Refined products generally need much less energy to digest than vegetables, fruits, wholemeal products or meat. 100 calories of chocolate will go straight into your system, while 100 calories of cabbage will take far more energy to chop and chew.

The number of calories shown on packaged food labels (where each calorie is a kilocalorie) is based on a calculation of the average calorie values of proteins, carbohydrates and fats called the Atwater system. Staggeringly, fibre is taken out of the equation, 'because it is not used by the body'.[43] The calorific values are based on calculations derived from burning foods in a tin can called a calorimeter. But if we are serious about 'calorie counting', the labels should be changed in order to reflect the differences in energy used to consume different foods – and should include fibre. The most significant difference may be in how the food is cooked, however – cooking releases food calories making them easier to digest. Processed foods make it even easier. We should update the calorie counts to take account of these factors.

The biggest distortion of 'it's all about calories' is in 'calorie-free' drinks. But the sweeteners replacing sugar in so-called calorie-free drinks encourage you to eat more.[44]

This is what we need to do:

1. Have an open, transparent debate, not a shouting match. Perhaps start with a conference or establish a government committee to examine the 'fats v carbs' issue, and make recommendations on the basis of that.

2. Carry out a global check to see if any dietary guidelines are bringing obesity rates down. If they are, let's look at them. Where they are not working, get rid of them.
3. Stop lumping sugars and fats together.[45]
4. Revisit the traffic-light labelling scheme, which signals red for danger – i.e. fat – and green for healthy – e.g. vegetables. Cheese should not be 'red'.
5. Redesign supermarkets. Highlight proper five-a-day, vegetable oils on salad, wholefoods, and include meat. Tear down those 'low-fat', 'healthy' signs over the cereal, crisp, biscuit, cake and bread aisles.
6. Tax *all* sugar. A start has been made with the sugary drinks tax. Tax sugar at source. That is the easiest and best way.
7. Find alternatives to sugar that are not artificial sweeteners and don't let HFCS in. With the abolition of internal quotas in the EU (HFCS), isoglucose production is expected to increase threefold. Asked if there were any health implications, EFSA said: 'Some recent short- and longer-term intervention studies … have shown that high fructose intakes (25 per cent of total energy) induce dyslipidaemia, insulin resistance and increased visceral adiposity … However, these effects are generally not observed at lower doses of fructose intake (about 40–50g/day in place of starch or sucrose).'[46]

So that's all right then.

We need different research asking different questions. Until now 'fat is good for you' has been dismissed as a fad – the Atkins diet, for example – and reducing sugar ridiculed – as happened to Yudkin 30 years ago. We need to deal with past mistakes, rather than pretending all is well. It isn't. People are confused as to what to do. While not all can be blamed on the guidelines, we need to admit they are not working.

One thing we can say, after Brexit. We won't be the fattest country in the EU anymore.

9
Pesticides

In this chapter, we'll look at one of the issues that will be devolved from EU decision-making – that of pesticide approval. This is an example of where the role of science will become more apparent as we make decisions previously taken in Europe, mainly by big players – the Brussels bureaucrats, agri-corporates and NGOs – opening up the prospect of many of us having more involvement in what is decided.

According to the NFU, the Water Framework Directive, which aims to achieve a 'good status' for all surface waters, offers the greatest opportunity for change. The NFU Public Health Officer says: 'We don't sell our water to the Europeans, so let's change it.' Currently the maximum permitted level of pesticides in water is 0.1µg/litre. The NFU complain that 'One hundred times as much arsenic is allowed in water as metaldehyde (slug killer) with no harm to humans. Why are we working to arbitrary levels? Let's set some new thresholds that are about safety rather than arbitrary.'[1]

Any trade deal the EU has agreed with third parties is on the basis of the current pesticides regulations. Any future trade relationship the UK has with the EU will be affected by these pesticide standards. Trade with countries outside the EU may provide an opportunity for greater regulatory flexibility.[2]

When we talk about pesticides, we refer to chemicals that kill pests. These include herbicides, which kill plants, insecticides, which kill insects, fungicides, which kill fungi, and so on. Two major pesticide issues, which have been raging for some time in the EU, will be coming our way.

INSECTICIDES

The big bug issue of the past few years involves insecticides that are called 'neonicotinoids' (neonics), modern chemicals based on the old very toxin, nicotine. Neonics were developed to replace more toxic insecticides. But it isn't their toxicity that has come under attack.

In 2008, Germany revoked the registration of one of them – Clothianidin, for use on seed corn – after an incident that resulted in the death of millions of nearby honey bees,[3] but reinstated its use when a better 'sticker' was used to keep the chemical on the seed. There are already strict controls on the use of neonics in France, Germany, Italy and Slovenia. Neonics seed treatment is banned in Italy, but plant-spraying is allowed.

But in late 2012, there was a new concern about these neonics,[4] as studies suggested they could change the *behaviour* of bees. The results suggested that current methods for regulating pesticides were inadequate because they considered only *lethal* doses of single pesticides. The present approval process looks at whether bees are killed, not whether they get lost. The study involved a combination of insecticides, was in the laboratory and involved bumble bees. Did this translate to honey bees in the field?

The discovery that neonics affected the behaviour of bees was a new impact, which we'd not considered. I'd been on the Advisory Committee on Pesticides (ACP), and behavioural matters weren't considered when deciding approval for use of pesticides.

Europe stepped in. EFSA were asked to look at the possible risk to honey bees of sub-lethal amounts of the pesticide in nectar, dust and the little droplets like dew on leaves. They identified a number of 'data gaps' that would have to be filled to allow further evaluation of the potential risks, including the risk to pollinators other than bees. As a result, the EU imposed a moratorium on the use of neonics until better data was available.

DEFRA came out with a report quoting 40 studies which said that 'any effects are likely to be small and not significant'.[5] However, government scientists warned that the report was flawed and the main study contaminated, as its controls had neonics.[6]

The UK's ACP discussed the issue in 2103 and concluded that the level of concern was sufficient to suggest a need for a review in line with EU regulations. The committee added that further analysis of the data would be required to identify whether or not there was a need for any regulatory action in the UK as a result of that review.[7]

Not everybody agreed. In 2014, Matt Ridley said the ban was supposed to protect bees, but damaged farmers' crops. He claimed crops of oilseed rape were dying off all over South-east England because of infestations of flea beetle.[8] However, a closer look showed patchy infestations, and suggested that other factors played a role in determining the level of damage, including seed-bed preparation, use of rotations (classic pest control), heavy land and sowing dates.[9]

My favourite study from 2015 was from the Proceedings of the Royal Society B,[10] which found the 'missing link' – why bees in the field might react differently to bees in the lab. Bees in the field counteract the immediate impact of neonicotinoid pesticides by producing more female forager workers at the expense of male drones, whose job is to breed. We know we cannot be 'anthropomorphic' (i.e. impose human ideas on insects), but the idea of losing lazy drones to produce more female workers may appeal to many. There are concerns down the line about queen fertilisation.[11]

In 2016, a study conducted over 18 years of over 60 wild bee varieties found 'the first evidence that sub-lethal impacts of Neonicotinoid exposure can be linked to large-scale population extinctions of wild bee species, with these effects being strongest for species that are known to forage on oilseed rape crops'.[12] This study will have a major impact on UK and EU decisions. Interestingly it also 'found that the application of foliar [plant spraying] applied insecticides had little or no negative consequences for population persistence of wild bees'.

In April 2017, the then Farm Minister George Eustace declined an NFU appeal to allow some farmers 'emergency' access to neonics for use on oilseed rape, on the advice of the Expert Committee on Pesticides (ECP – the body that replaced ACP). The ECP said industry hadn't honoured commitments to develop more targeted pesticide use and carry out monitoring work that they had agreed to do in 2015.[13] In May 2017, Rothamsted Research Station issued a 'Position Statement'[14] saying that the EU restrictions were based on 'contested science' making it difficult to maintain production of many crops in the UK.

If I were still on the ACP, I would say we should ban neonic seed treatment. This sort of treatment is called 'prophylactic' – i.e. acting 'just in case'. The parallel is the regular use of antibiotics in cattle in America – whereas we only use antibiotics when needed. Regular prophylactic use leads to quicker build-up of resistance in bugs.

However, that key study from Royal Society B leads the way to allowing foliar sprays – when necessary – if an outbreak of flea beetle occurs. This should be done only as a last resort and following proper implementation of an integrated pest management programme,[15] and the NFU carrying out the target trials the ECP required (above). At present, integrated

Pest Management is required by the EU Directive on Sustainable Use of Pesticides (SUP).[16] The Cameron government produced a National Action Plan[17] in line with EU law. There are also International Guidelines on IPM – GlobalGAP[18] – and the NFU has introduced a guide for IPM[19] as part of the Voluntary Initiative.[20]

IPM means that farmers/agronomists will need to use rotations, build better seed-beds, keep an eye out for the pest – and then use properly qualified applicators. They should also be investing in long-term research. Three-year funding won't work to examine other ways to control pests like flea beetle. I was told as a student that it took 15 years to study an insect pest properly. In the meantime, there needs to be a lot more advice on chopping and changing crops, lengthening rotations to break up pest cycles, earlier sowing and better preparation of seed-beds.

The UK enacted the Sustainable Use of Pesticides Directive pretty well, so shouldn't rush to junk it. We have a well-organised system for training sprayers provided by regulated private-sector organisations and institutions such as agricultural colleges. I think HSE's Agricultural Industry Advisory Committee (AIAC),[21] and its subcommittee, CHEMAG, can take a lot of credit for that requirement, prompted by Unite over many years.

Post Brexit, there may well be attempts to water down the pesticide safety levels, as we saw at the start of this chapter. However, having different regulatory systems makes it more difficult for global companies to bother with UK authorisations. We also have a good track record in setting standards beyond the law, as seen among retailers, through to our Chemical Regulation Directorate which has a well-respected record for generating evidence and expertise.[22] We should make use of that, not rubbish it.

HERBICIDES

The first stage of the process of pesticide approval is by the European Food Safety Authority (EFSA) and the European Commission. They decide whether the 'active ingredient' passes a long list of tests, costing the companies around £100 million. An 'active ingredient' is a chemical like glyphosate, good at killing plants. Then – through our own approval process – the 'formulation' of the active ingredient must be approved, in this case, a formulation known as 'Roundup'. This is the trade name you see on packets in the garden centre: in very small writing you will see the word 'glyphosate'.

Glyphosate is under the spotlight. It is the most widely-used weed-killer in the world, sprayed on at least 100m acres worldwide. The NFU in their 2017 manifesto 'Brexit and Beyond: 5 Ingredients for Success'[23] say:' The UK Government should commit to a science-led approach with fit-for-purpose legislation to approve safe and effective tools such as glyphosate and neonicotinoids.' How we handle the dialectic between 'science-led' and 'fit-for-purpose' will tell us a lot about how we manage our crops in the future.

The International Agency for Research on Cancer (IARC, part of WHO), a prestigious body, decided that glyphosate was 'probably' a carcinogen, not merely 'possibly' carcinogenic. The agency published a monograph (112) based on their scientists' advice which said that glyphosate should be reclassified in the WHO rankings from a 2B chemical (possibly cancer-forming) to 2A (probably cancer-forming) chemical. Monsanto asked for a retraction, saying it was based on junk science. EFSA – the same committee that had pointed to the gaps in neonic evidence – was asked to look at glyphosate. They concluded that 'Glyphosate is unlikely to pose a carcinogenic hazard to humans and the evidence does not support classification with regard to its carcinogenic potential.'[24] Accusations of bias flew everywhere.

NGOs campaigned to have glyphosate banned. The fact that it was connected to Monsanto's favourite GM Roundup Ready crops added to the fervour. The EU authorities were accused of being in the hands of the corporates. The more you looked, the more cover-ups and accusations appeared – but little actual evidence. However, it is noticeable that nobody campaigns to ban hairdressing, burning wood in the home, glass manufacturing, red meat, shift work or high temperature frying – all also classed as 2A activities by the IARC.[25]

The two organisations, IARC and EFSA, employ different approaches to research. The IARC uses pretty transparent studies of glyphosate containing herbicides (formulation), while EFSA uses peer-reviewed studies based on industry-sponsored research into the active ingredient.[26] IARC allows assessors to include their own research papers, EFSA does not. The weakness of both – as IARC pointed out – was that while glyphosate is the most frequently used herbicide in Europe, there is little information available on occupational or community exposure to it. In other words, nobody is monitoring workers and others in contact with the chemical. Monitoring trumps modelling any day.

There were challenges to the re-licensing of glyphosate, which was due for renewal in the EU. In the event the licence was extended for only 18 months, instead of 15 years, starting from June 2016.[27] A further report from EFSA is expected. Further licensing will soon be in the UK's own lap.

There is concern in the US that the safety levels there are based on old data and should be re-assessed.[28] That may well be the case – in the US, where levels are much higher than in the EU. In the EU, the ADI (Acceptable Daily Intake, see below) is currently set at 0.3mg/kg bw/day, whereas in the USA it is 1.75mg/kg bw/day. That is quite a difference, and shows how 'protective' the EU is.

We have spent years fighting for improved spray methods. In Unite, on the Health and Safety Executive, we argued we did now want a 'cover up', where workers had to wear protective clothing, but regulations to prevent exposures happening. But control conditions are not the same elsewhere, where they are often not a patch on those in the UK. Again, we export our environmental and health impacts abroad. The alternatives to glyphosate are hoeing, covering crops, or using another weed-killer like paraquat. No thanks to the latter. While writing this book, I heard from farmworkers in Asia that paraquat is still being sold as a liquid: it looks like cola and one mouthful kills, accounting for too many deaths across the globe.

I would argue that glyphosate should be restricted from pre-harvest treatment (see below) and that its license should be regularly reviewed – say, every three years. There should be a monitoring scheme set up to find out the extent to which workers and others in contact with the chemical are actually exposed to it, to enable any new information to be factored in, but also to allow a period of time in which manufacturers could be required to do additional research, on environmental impacts, for example.

For me the issue isn't so much about the toxicity of glyphosate, but more about the environmental impacts. Look at what it says it does on the tin: it kills weeds – i.e. plants. Monsanto are not going to argue with that. We do not know what the consequences of killing all those plants are, all those trillions of little carbon-capture storage units.

DIGGING UP THE DIRT

My PhD, in the early 1970s, was designed to try to find how herbicides affected populations of soil animals. I want to tell this particular story to help cast light on one of the greatest issues of our day – carbon control.

At the time, many of us were strongly influenced by Rachel Carson's *Silent Spring*, seen as the beginning of the environmental movement. Ecologists were haunted by pesticides called 'Drins' (Dieldrin, Aldrin etc.), which were thought to kill off birds of prey through the contaminated food chain. My thesis started during the first UN Environmental Conference in 1972. At that conference, (toxic) pollution was mentioned four times, but there was no hint of global warming.

The field of soil ecology was new to me. Having studied agricultural entomology, I was expecting to find lots of insects in the soil. Yet I found that most of the 'soil mesofauna' – those creatures smaller than earthworms, but bigger than bacteria – are not insects. In the intervening period, I have come to reassess the significance of soil ecology. It is all about 'pesticides' and 'paradigms'.

I had been working in the root laboratories at East Malling Research Station,[29] where you can see the creatures live in the soil from the vantage of an underground laboratory. We collected soil animals by taking cores of soil with an 'auger' and extracting the animals by shining lights on the overturned soil core to drive the animals down into collection vessels.[30] It is a bit like shaking a tree and collecting everything that drops out, then trying to work out how it all fits together. My thesis was to test the effects of weed-killers on them.

I carried out my research at Wye College, London University, then probably the most prestigious agricultural college in the world.[31] I discussed with my supervisor which herbicides I should study. He said: 'don't bother with 245-T, it will be banned by the time you finish.' This was based on what we already knew about the effect it was having on soldiers in Vietnam. It was never banned, although later I – along with colleagues in the farmworkers' union – spent many years trying to do so.[32]

I linked up a lot with my mates in the Soil Zoology Unit at Rothamsted Research Station, partly as an excuse to go drinking with them on a Friday, but also to discuss what I was doing. They helped me determine that what I needed to do was to try to establish whether the weed-killers had any toxic effects – these 'direct' chemical effects rather than 'indirect' effects, caused by the loss of herbage. We were very concerned about toxicity – poisoning – then.

I planted out a trial area with 40 small plots, 4 each for 10 treatments (including controls). The patch of land was right in front of the Principal's house, so I had required special permission to dig it up (see Photo 9). I built a massive extraction unit consisting of 60 'Tullgren' funnels, down which the small soil animals fell into bottles of alcohol. I counted over half a million creatures in three years, identifying them according to 70–80 species. That made for a lot of statistics.

Photo 9 Charlie Clutterbuck's testing weedkillers field trial plot at Wye College in 1970.

I spent ages trying to identify possible toxic effects, comparing each herbicide versus their controls using a great new invention called the London University mainframe computer. The computer output took 2 hours for each monthly run. Despite all this, there were few clear toxic effects – other than that caused by loss of herbage. I had weeded out a few toxic effects, but they could easily have been random effects. So I was worried.

I duly finished and presented the thesis. At the time, I thought I had failed, because I had not found any toxic effects, whereas the external examiner felt that I had succeeded – in showing that there *weren't* any toxic effects.

Over 30 years later, while serving my time on the ACP, it dawned on me. We were discussing the difference between pesticide approval and GM approval. As I spell out in the chapter on GMOs, these go through an extra requirement – of checking for indirect effects as well as direct effects. Then it hit me. Why didn't I look at both direct and indirect impacts of herbicides all those years ago? It doesn't matter to dead soil animals whether the effect is direct (toxic) or indirect (dead herbage). No wonder I couldn't find any direct effects all those years ago. The main effects are indirect, as that is the whole purpose of herbicides – to kill plants.

Glyphosate is usually used on ploughed fields to keep out weeds – particularly black grass – to help the desired crop grow. I discussed in Chapter 6 the loss of soil organic matter in the last 30 years. This loss is often put down to ploughing. But it is unlikely that ploughing alone could explain the loss, as there is no sign of an increase in ploughing during that time. Yet there has been an increase in total herbicide usage over the same period. Glyphosate use in UK farming has increased by 400 per cent in the last 20 years, so it is more likely to be a possible cause of soil organic matter reductions. More than 4m acres of land were treated with glyphosate in England and Wales in 2014. This could account for a large increase in CO_2 being left in the air, rather than being captured by the weeds and their herbage returned to the earth – and stored as soil carbon and soil life. Perhaps in future we will regard our weeds as climate friendly. If we could work out the contribution of weed-killers to global warming, then we could reward those farmers choosing not to use herbicides or planting low-growing plants under the main crop, and thereby becoming a carbon-capture and storage-field factory.

Currently, a much greater area of crop is treated with glyphosate to aid harvesting (pre-harvest) than for weed control. Benefits have been recorded in wheat and barley crops where there were sufficient weeds to delay and/or slow harvesting operations. It should be possible to calculate the loss of carbon from the soil due to loss of weeds, and what that means in terms of extra carbon dioxide in the atmosphere. But I can't see Monsanto funding it. Here is an example of the importance of public-interest research for the public good, rather than research for private profit.

The government has consistently said it would vote in favour of reauthorising glyphosate, should the EU ban it, which it hasn't. According to Adam Speed, a spokesperson for the UK Crop Protection Association, it is another example of 'the EU ignoring science'.[33] As Mandy Rice-Davis famously said: 'He would wouldn't he?' I suggest we get the UK government to put some money into researching how much carbon can be saved by not using herbicides and the opportunities for mitigating carbon dioxide emissions by encouraging low-growing plants.

POST BREXIT

When it comes to pesticide control after Brexit, I predict three major moves the UK government will make:

Move from a 'hazards' approach to one that is 'risk based':
The regulation of pesticides is unlikely to be top of the Brexit agenda, but should be part of the Repeal Bill. Then we can talk about reform in the longer term. There is an important distinction between EU 'directives' and 'regulations'. People are pouring over the UK legislation to find out what is 'operable' (i.e. can be 'lifted and shifted') and what is too closely linked to EU institutions (inoperable). Pesticide laws provide an example of the complexities of deciding whether we can transfer the existing law straight into new UK law. Pesticide regulations can be lifted and shifted, but the directives are harder to interpret. Directives like the Sustainable Use of Pesticides will need to be converted into UK law by Acts of Parliament. I can hear the howls now, as it comes to the Commons.

Many will talk about a more flexible approach – called 'risk assessment'. The EU law laying down Maximum Residue Levels (MRLs) looks like an example of operable law. The EU lays down these pesticide levels in food as a part of its legislature.[34] The MRLs represent the 'hazards' approach, in that a particular level is determined that should not be exceeded. The EU approach is to decide criteria for cut-off points for pesticide control.

MRLs are derived from the NOEL, not the one sung about at Christmas, but the No Observable Effect Level. These levels are an example of a 'hazards' approach – the level is set based on the inherent toxicity of the chemical. The NOEL is set where nobody has been found to be ill as a result of that level of exposure. Retailers take MRLs seriously and will check their foodstuffs regularly against the permitted levels. Another level derived from NOELs concerns Acceptable Daily Intakes (ADIs), which set out what people and workers can be exposed to in a day with a degree of confidence about their safety. I thought that the international body of the WHO and FAO, the *CODEX Alimentarius,* decided on pesticide levels, like MRLs.[35] So I was surprised that the US and EU ADIs for glyphosate, seen above, differ so much.

There is a genuine concern that people just pick on a hazard trigger when they fancy and shout 'ban it', as happened when glyphosate was classified by IARC as a 2A compound. There needs to be consistency. To bring clarity, a group of us advised the Cooperative Group (CWS) to help deal with pesticides as part of their Responsible Retail initiative. The aim was to develop and apply a 'hazard-trigger algorithm' to all pesticides in routine use in farming throughout the world, in order to prioritise those that might need controls that were tighter than the legal requirement. A two-tier hazard-trigger algorithm, based on internationally recognised values, was applied to over 800 pesticides, and compounds that 'tripped' a hazard trigger were assigned to some level of restricted status. 132 were placed on a proposed prohibited-use list and 325 on a monitored-use list.[36]

The international values used included the ADIs, the WHO carcinogenic class (used above), acute toxicity, along with soil and water persistence. In the case of occupational exposure, the triggers are based on widely-accepted standards of exposure. We said the existence of a Maximum Exposure Level (MEL) for a pesticide (only the more dangerous chemicals have MELs) put it on the proposed prohibited list, and an Occupational Exposure Level (OEL) 1mg/m[37] put it on the monitored list. This is similar to the Code of Practice for using Plant Protection Products (DEFRA[37]) that suggests exposure levels can *trigger* health monitoring. This approach leaves it open to select which internationally recognised values to use, what levels to select, and the consequences in terms of controls. This seems a sensible and scientific approach that is easy and consistent to use.

However, the UK will prefer a 'risk-assessment' approach rather than a 'hazards' approach. This means that the hazards are identified, then the likely exposure assessed and factored into a risk assessment – i.e. the likelihood of damage. It may be toxic, but if the exposure is low, we don't have to bother. Britain used to say the same about air pollution – why bother, if it blows away. The 'risk-assessment' approach will prevail as we exit.

All our Health and Safety law is based on 'risk assessment'. Many of you may well be sick of the term. I used to tutor risk assessment with trade union reps, explaining that you assess the risk from a given hazard. First identify the nature of the hazard then determine the likelihood of it reaching you, then prioritise the controls based on that. We won those rights and responsibilities in the early 1990s to risk assess everything at work. The government could easily say that we should adopt the same approach for pesticides.

Drop the Precautionary Principle:
The Precautionary Principle was agreed at the Environment Conference in Rio in 1992, Principle 15 says: 'In order to protect the environment, the precautionary approach shall be widely applied by States according to their capabilities. Where there are threats of serious or irreversible damage, lack of full scientific certainty shall not be used as a reason for postponing cost-effective measures to prevent environmental degradation', i.e. 'where there is a serious risk, we shouldn't wait for the science to stack up'. It was intended for use on matters like global warming or depletion of the ozone layer.

However the EU has used it for GMOs and for neonics, much to the annoyance of many in the UK (and the US). The government say they prefer an evidence-based approach. Everybody agrees with that. But the trouble with that is people like to choose their own evidence.

Carry out the whole pesticides approval process ourselves:
According to AHDB

> Brexit offers an opportunity to join up the standard-setting and registration processes, which are currently shared between the European Food Safety Authority (EFSA) and the Commission, to something similar to that used in the USA where the Environmental Protection Agency (EPA) has responsibility for the setting of standards, assessing compliance and final approval.[38]

At present the EU body EFSA and the European Commission, who do not always see eye to eye, determine whether an active ingredient is approved. Once approved, it comes to the UK, where the ECP advises ministers on the safety of any particular formulation. In the future, I can see the ECP will determine the fate of active ingredients as well as the formulations.

I suggest it is also an opportunity to create a new government pesticide agency – one to monitor what happens in the fields after approval. We have looked at two groups of chemicals, neonics and glyphosate, where accusations of bad science have been much in evidence. We will need better evidence in the future, when these chemicals will be 'in our control'. We may create innovative ways to control them, but we will need to monitor how effective they are.

MARCH FOR SCIENCE

We have seen, in this chapter, how people claim to have 'science on their side'. There is increasing concern about the lack of respect for science. This is most evident among the 'climate change deniers' and vaccine sceptics. In response, there is a global movement developing called 'March for Science'.

Many people who claim to be scientific accuse others of being corrupted by corporate money. I am the first to recognise that society influences science. What questions are asked are determined by the prevailing ideas, and answered by the funding mechanisms. To get funds you must mention global warming or food security.

While I was working at the British Society for Social Responsibility in Science this was a hot topic. There were those – prominently Robert M. Young – who argued that all science was just a reflection of society, that all science was a social construct. Others, in particular the Roses – Hilary and Steven – recognised that while society did influence science, they nevertheless considered there are demonstrable replicable truths underneath. Pesticides and fertilisers were developed in wartime – but they work.

Science isn't a shouting match. We have to look at and evaluate the evidence. Difficulties arise when the evidence presents you with something you don't expect or don't want. It happened with my own research. After digesting the evidence, we should come to conclusions, but keeping an open mind, as new evidence emerges. It is the responsible thing to do – to bite into something, chew it over and slowly digest it.

How we deal with these pesticides says a lot about the way in which we deal with science issues. We do not want to give any succour to climate change deniers and the like. Science isn't just a matter of throwing 'facts' about, as many are now doing. Just because a scientist says something, doesn't mean it is true. There are always conflicting views in science, and long may that remain the case.

The first step is to make a theory – like the one I did in Chapter 6, regarding the birth of the Earth. Then you have to set out to disprove that theory. We do this by setting up tests. And then some more tests. From these tests, and other sources, we build up evidence. The evidence acquired is never perfect, so we have to be critical about all evidence presented, and then weigh it all up. In toxicology, the results may apply to animals, but be irrelevant to humans. The other science dealing with hazards – epidemiology – looks at patterns of disease, but can rarely isolate populations clearly for study. For instance, regarding glyphosate, it would be impossible and unethical to mount a trial to identify an unknown cancer by following two groups of people (one exposed and one not) for many years. Both sciences are constrained by money, yet we rely on them. All that evidence then requires critical appraisal, and often I find that there are layers of truth; like peeling an onion, you keep finding something deeper below.

Over the years, the ways we have looked at the environment – and thus the science we wish to explore – has changed. In the early 1970s, we were concerned about toxins – chemicals poisoning ourselves and the planet. Hence the focus on the toxicity of herbicides in my own research, but also the toxicity of insecticides – like the neonics. Generally these chemicals are safer than they were. And in the last 20 years, our environmental concerns have been more to do with global warming, which is not caused by toxins (like carbon monoxide) but by everyday chemicals (like carbon dioxide), in overdrive. Similarly our main concern with food now is obesity, which is caused by natural compounds in overdrive – nothing to do with toxins, like pesticides, in food.

Let's look at the role of science in that other 'great debate' in food and farming – GMOs.

10
Genetically Modified Organisms (GMOs)

This chapter aims to explore some of the issues that will arise as a result of coming out of the EU and thereby allowing the UK to decide what to do with GM foods. While they have been the subject of a lot of argument, perhaps now is the time for a rational debate. This issue will probably generate the most heat after Brexit, so it is worth a closer look.

Princess Anne (Honorary President of the Oxford Farming Conference) spoke to *Farming Today* in March 2017, saying 'surely if we're going to be better at producing food for the right value, then we have to accept genetic technology is going be part of that ... I could see it growing on my land.' The media reminded her that she contradicted what her brother, Prince Charles, had said in the 1990s – that GMO 'takes us into areas that should be left to God. We should not be meddling with the building blocks of life in this way.'[1]

All hell broke loose, after Charles's remarks. Two supermarkets were targeted. Sainsbury's and Safeway's had for three years sold two forms of their own-brand tomato paste, the cheaper one declaring GM. Nobody had any problems with the choice offered, until Prince Charles stepped in. Then those two supermarkets had to clear away not only the tomato paste, but also anything with tomato paste in it, like pizzas and sauces.[2]

Up till now, the approval of GM crops has been in the hands of the EU. To explain the EU position on GM would require a PhD thesis, but here is a flavour. In April 2017, Members of the European Parliament voted against the Commission (Brussels bureaucrats') plans to authorise imports of food and feed products derived from or containing a GM maize variety from the Syngenta company. MEPs called for reform of the EU's GM authorisation procedure, saying the Commission had ignored a three-month consultation. The Commission said its plan was based on the advice of its scientific watchdog EFSA, who gave a favourable opinion in August 2016. There was much more about which procedures should be followed. Explaining how the EU GM policy works is a metaphor for the EU itself.

Now we are coming out of the EU, a new approval process will be up for debate, although debate is hardly the word to describe what is likely to happen. Much energy will be consumed arguing over it. This is why I will look at this issue in more detail in this chapter.

While I was on the government's Advisory Committee on Pesticides, we were asked to compare the GM approval process with pesticides approval. We noted the clear distinction that pesticide approval looks at possible *direct* effects of the chemical on everything, whereas GM approval requires both this and *indirect* effects. An example of an indirect effect would be what Peter Melchett, Policy Director of the Soil Association, describes in his article 'Seven Sins of Science'.[3] 'These products … alter the relationship between farmers and seed producers, preventing farmers saving their own seed.' *Indirect* effects like this can refer to almost any aspect of growing, so it explains – in part – why there is only one GM crop approved in Europe.

TOMATO CASE STUDY

At this point the educator in me comes out. There are many issues connected with GM, so let's try and tease out some of the main concerns, by comparing two forms of new purple/reddish brown tomatoes. You decide which you would prefer to be developed.

Purple Tomato

In *Current Biology*, Cathie Martin and colleagues[4] studied tomatoes enriched in anthocyanin, a natural pigment that confers high antioxidant capacity. Many 'celebrities' swear by these anthocyanins, found in cranberries, bilberries, etc., leading to the term 'super fruits'. They also slow down the ripening process and delay rotting, leading to a long shelf life and full flavour. The purple tomatoes are also less susceptible to one of the most important post-harvest diseases, grey mould, reducing the need for fungicide use.

This purple tomato plant, with anthocyanin throughout the tomato, not just in the epidermal layers, has been developed at the John Innes Research Centre.[5] Their spin-off, Norfolk Plant Sciences, pays for the patents,[6] using genetic modification (GM) techniques. A gene from a snapdragon plant is inserted into the tomato chromosome. 'Working with GM tomatoes that are different to normal fruit only by the addition of a specific compound, allows us to pinpoint exactly how to breed in valuable traits', said Professor Cathie Martin from the John Innes Centre. If you plant the seeds, the ensuing plants grow true.

The Big Barn blog asks the important question 'Are GM tomatoes healthy?' They say: 'I am not a scientist but doubt anyone can be 100 per cent sure that when transferring a tiny gene that something else may also be transferred undetected. And that as tomatoes are members of the deadly nightshade family, a poisonous gene may be awakened. And that it would be a lot easier to add the anti-oxidant anthocyanin to ketchup and pizza sauce separately.'[7]

I think Heinz could add anthocyanin to ketchup, but it seems more natural to grow it. I find it interesting that 'deadly nightshade' is mentioned, as the same concerns were raised when people were protesting against potatoes in the early 1600s. There are other GM parallels with potato introduction. Many Protestants would not plant potatoes, as they hadn't been mentioned in the bible. The other main source of prejudice against the efficient tuber was that they were part of the nightshade family, which includes several poisonous members. Nightshade was associated with witches, famous here in Lancashire due to the Pendle Witch Trial in 1612, which had more to do with landownership than witches' brews.[8]

Solanine in tubers is a poison, and may have caused rashes that became linked with leprosy. Despite recommendations from the Royal Society, there was still fear. In the late 1770s potatoes became more popular when Louis XV1 and Marie Antoinette wore potato flowers as buttonholes. In 1774, Frederick in Russia issued an order to grow potatoes to protect against famine. The town of Kolberg replied: 'The things have neither smell nor taste, not even the dogs will eat them, so what use are they to us?' So Frederick planted a royal field of potato plants and stationed a heavy guard to protect the field from thieves. Nearby peasants assumed that anything worth guarding was worth stealing, and so sneaked into the field and snatched the plants for their home gardens. The message spread.

Instead of growing our own purple tomatoes, we imported 2, 000 litres of purple tomato juice from Canada in 2014.[9]

Kumato

The Kumato is the trade name given to a variety of tomato developed in Spain called 'Olmeca'. It is grown in Spain, France, Belgium, the Netherlands, Switzerland, Greece, Turkey and Canada, where it is called Russo Bruno. The Kumato is a standard size, reddish brown and sweeter than typical tomatoes because of their high fructose content, according to Syngenta. Syngenta is protective of their 'super-scientific' crop, and have patented the Kumato. It can only be grown by specially selected growers.[10]

This plant has been developed for eating pleasure and is *not* genetically engineered, but a *hybrid* of two lines of tomato grown separately from each other. This means two lines of separate varieties are crossed in the traditional way. The planted seeds will not grow plants identical to the parent, but will have increased vigour. Syngenta say that one of the lines comes from Galapagos – hence the black quality. Their seeds cannot be purchased by the general public. Syngenta state that they will never make Kumato seeds available to the general public, as the Kumato is grown under a concept known as a 'club variety'.[11] Syngenta sells seeds only to licensed growers who go through a rigorous selection process, and participation is by invitation only. Syngenta maintains ownership of the variety throughout the entire chain from breeding to marketing. Selected growers must agree to follow specified cultivation protocols and pay fees for licenses per acre of greenhouse, the cost of the seeds, and royalties based on the volume of tomatoes produced. Typically, Syngenta licenses only one large vertically-integrated greenhouse producer per country. As we will see, Syngenta has now been bought by the Chinese.

It was such 'hybrid vigour' that drove the Green Revolution – in particular hybrid maize in the Americas and hybrid rice in Asia. You need large growing areas to develop the separate strains before crossing, so this was the time the corporates took over the seed markets from smaller breeders. The crops used more water and more chemicals, so it was the chemical companies who saw the opportunity and became botanical. Monsanto was one and Syngenta another.[12]

PRIVATE

The first GM crops were developed by the major corporates, but as the cost of GM technologies has plummeted, it is now in the hands of the public sector, and in some cases in the hands of individuals working in their kitchen.[13] How does this change the issues?

The big agri-corporates are undergoing major reorganisation as they search for more profits. In the last year, the dominant agribusinesses have gone from six to three. Syngenta was bought by China, Bayer bought Monsanto, and Dow and Dupont merged. These three firms control 60 per cent of commercial seed and more than 75 per cent of agrochemical markets

The biggest deal was Syngenta, which was acquired for $43bn by the Chinese state.[14] The acquisition includes the research station at Jealott's Hill in Reading, which I knew as ICI's, which will now be in Chinese hands. ICI split into two, one part Astrazeneca, which later merged with Novartis to create Syngenta. Novartis and AstraZeneca had wanted to establish themselves as 'life sciences' companies to exploit potential synergies between their pharmaceutical, chemical and agricultural sectors. Both invested heavily in acquiring seed and biotechnology companies, but suffered in part from the GM backlash. Syngenta became the world's largest agribusiness.

Syngenta says the Chinese have bought their Good Growth Plan.[15] This pinpoints six commitments – to make crops more efficiently, rescue more farmland, help biodiversity, empower smallholders, stay safe, look after every worker.[16] Syngenta is over twice the size of its rival Monsanto[17] and turned down their higher bid. That may be because they were reluctant to help Monsanto with sales of Roundup – increasingly in need of Syngenta know-how at overcoming the build up of weed-resistance. The Chinese have bought not only the chemicals, and the seeds, but also a lot of 'knowhow'. They need it, as they have 21 per cent of the world's population with just 9 per cent of its arable land; a lot of young people have moved off that land, and there are many environmental challenges.[18] China will own several important patents including strobilurin fungicides, the insecticide abamectin and triazine herbicides, along with paraquat. The US fears that China will now block US GM imports. For Syngenta, it gives them muscle against Monsanto and increases their potential output of patent-protected seeds.[19]

The merger that is attracting most attention is that between Bayer and Monsanto. Previously these two companies joined forces to make 'Agent Orange' which was sprayed over Vietnam during the war to clear forests, causing birth defects among soldiers and Vietnamese.[20] Bayer plans to take over Monsanto for $66bn and it looks as if it has the approval of President Trump. Environmental campaigners lobbied the EU Competition Commissioner in March 2017 to try to block the 'marriage made in hell'.[21] The companies say they will spend more on research, but many fear the take-over will lock farmers in even more – thus threatening Trump's base of support. Bayer's base is in pharmaceuticals, Monsanto's in GM. A merger would boost their biotech division.

The Dow Dupont merger creates the second biggest chemical company in the world. The US Farmers Union are alarmed at consolidation in agriculture 'that has led to less competition, stifled innovation, higher prices and job loss in rural America'. While the corporates talk up 'more choice', the reality is that dozens of small competitors have been bought out, so driving up seed prices.

PUBLIC

GM in the public sector plays a different role from that in the corporate sector. Public research looks at desirable traits, rather than those that only bring increased profit and control. So a whole range of issues is addressed.

Whiffy Wheat was grown at Rothamsted Research Station.[22] It is a GM variant that resists aphid attacks and also encourages their predators, ladybirds. GM was chosen after other approaches did not provide or deliver the repellent odour. If successful, this variety will be in the public sector and will thus have the intellectual property rights protected – i.e. they cannot be stolen, bought or privatised. While it worked in the lab, it was not so successful out of doors. Presumably the whiff was blown away. This is another example of why we need land-research stations – not just life science labs.

Plants containing fish oil have been developed by teams at Rothamsted and the University of Stirling. Fish oils are considered healthy fats, but there are environmental concerns about depleting fish stocks or farming fish. Plants containing omega 3 oil are being developed, so vegetable matter could replace wild fish or farmed salmon as providers of much needed oils.[23] The two teams made omega 3 fish oils in GM Camelina plants, a member of the brassica family.

A GM late blight-resistant potato has been developed by Sainsbury Lab and Teagasc.[24] British scientists have developed GM potatoes that are resistant to the vegetable's biggest threat – blight. A three-year trial has shown that these potatoes can thrive despite being exposed to late-onset blight. Mind you, SPUDS (Sustainable Potatoes United Development Study) give away non-GM blight-resistant potatoes in Ireland that seem to work as well.[25]

GM potatoes that release less acrylamide have been accepted by the US Department of Agriculture.[26] Acrylamide is a chemical potentially carcinogenic to humans – produced when cooking French fries and potato chips. Although we have never got worked up about it in this country,[27] great efforts have been made in Germany to cook chips at lower temperatures.[28]

Although developed by Rockefeller and Bill Gates, Golden Rice has been made public.[29] This GM rice delivers beta carotene to improve the health of the malnourished poor. Deficiencies of carotene blind a quarter of a million children a year. While not the sole answer, this rice could help tens of thousands of children. *Scientific American*[30] says Golden Rice opponents should be held accountable. They estimate that 'the delayed application of Golden Rice, since its creation in 2002, in India alone has cost over 1.4 million life years'. Opposition to the rice says that it costs more, negating its benefit for the poor, and the nutrient level isn't sufficient to have any effect and may not deliver enough carotene. Greenpeace and some Filipino unions say it is just a sales pitch for the government to get more tax. These are examples of the indirect effects of GM. They are valid, but do they justify banning GMOs? This sort of debate will be important in a post-Brexit UK.

We need a discussion about who makes money from developments in the public sector. Instead of giving away the products, UK research should maintain ownership and sell them, so the profits can be ploughed back into further research on a range of sustainable food projects. The gene for weed-killer resistance is from a bacterium in the soil. If Monsanto paid 0.00000000001 cents or pence for every pound of GM weed-killer resistant seeds they sold, that could go to better sustain the soil. There are a few places where the state is developing GM as part of their sustainability agendas, including Cuban 'Socialist GMOs'[31] and China Communist GM, where poor farmers are growing much more GM than small farmers elsewhere.[32] However, the public didn't take to India's first GM crop.[33]

CONCERNS

I was sceptical about GM, thinking it was pushed through without much debate. I really didn't like the idea of using more weed-killers, or anything encouraging increased use of chemicals. So, I had to cheer when another GM crop came along that reduced the use of insecticides. Insecticides are more dangerous to us than weed-killers, as chemicals killing insects hit nerve cells that are biochemically very similar to our own.

What I had against both groups of GM was that the genetic modifications were owned by a few massive corporates. These particular corporates have a specific interest in chemicals: they had been chemical giants before buying out the seed companies. They knew that the long-term value lay in the seeds. These companies used the chemical way of patenting, to patent bits of life. This was unthinkable a generation before. In the 1930s our government would have said that these patents should be owned by the state – as a public good.

The issue is whether all the misgivings warrant bans. Here are the main concerns:

1. There is a clear case that the corporates are trying to take greater control over farming in order to keep extracting profits.[34] Their research goes into a few big crops in the world – soya, rice, maize, wheat, sugar, then coffee and cotton. The same few companies that produce GM also produce the pesticides and other chemicals to be used with the GM. The US companies want plant varieties to be patented in the same way as chemicals. The big corporations who developed the GM trait often claim it now belongs to them, ignoring the several thousand years of breeding that went in beforehand. There are multilateral processes in place, including the Union of Protection of Varieties of Plants (UPOV)[35] to protect plant-breeders, but some patents go way beyond that.[36]

Any post-Brexit UK-US Free Trade Agreement is likely to involve this ownership issue. US corporates want 'patents' to protect their plants rather than Plant Breeder Rights (PBRs), which stop anybody taking the breeders' seeds and selling them elsewhere, but don't stop the grower replanting the seed or developing new varieties. Patents do not allow that, but are not confined to GM. In the UK in spring 2013, many barley farmers wanted to plant their own barley seed, as their winter-sown barley had failed, frozen in the soil. They were not allowed to use their own seed from the previous year to reseed as spring barley, as the non-GM seeds were patented.

2. There are concerns in developing countries that their scientists and producers can only obtain genes and seeds from foreign corporates, and that biotechnological research does not focus on the local varieties of crops that are important to the world's poor farmers. It has always

been that way. Local crops grown using local skills are not the ones capital wants – they want to control money-making crops. *Biotech Ambassadors,*[37] in the Food and Water Watch report, shows how the US government pushes GM all over the world on behalf of its corporates.

3. We heard a lot about the 'Terminator Gene' – a gene to stop the seeds of the crop growing. But none has been found. Yet 'hybrid' seeds do just that. As we saw with the Syngenta Kumato, this involves two strains of the crop being crossed to produce a crop with 'hybrid vigour'. However, when you plant seeds from that crop, they don't perform very well. By definition, farmers cannot replicate the hybrids in the next generation. Chemical companies used this property to develop the power they now have over present crops – before GM was invented.

4. The *Daily Mail* came up with phrase 'Frankenstein food', and it stuck. Yet there is no evidence of any food poisoning. I checked out the trials that claim otherwise and looked at 16 studies quoted by those opposing GM. These include pigs fed with GM corn, Seralini's rat-feeding, detection of transgenic plant DNA in milk, a Russian study feeding Roundup Ready soya to Italian mice, FSA human feeding trial, Newcastle University looking for trans genes escaping the gut, Monsanto trials, a Danish pig farmer, Aventis's chicken trials with their herbicide-tolerant Chardon, Australian researchers and anti-pea weevil gene. Only in this last trial were there any ill-effects that could be put down to GM – it caused an allergic reaction. It was not approved for use. Imagine if we removed all foods that caused allergens.[38] 88 per cent of scientists consider that GM foods are safe to eat.[39]

There are hundreds of millions of hectares of GM stuff grown, and millions of people looking out for any signs, and after 20 years there is nothing. Meanwhile, workers at a well-known horticultural centre have contracted lung disease from turning 'organic' compost, caused by *Aspergillus* fungal spores.

This brings into question the continued use of the Precautionary Principle to justify the EU GM approval regime. We saw in Chapter 9 that it relates to 'serious risk'. It is likely the farming industry will try to do away with it, which would be a shame as it is useful. We have to ask ourselves whether GM offers serious risk and good use of the principle, or whether the principle should be time-limited. If no serious risks have been found after 20 years, we should move on.

5. GM will 'contaminate' pristine ground nearby. Organic farmers with land adjacent to land on which GM crops are growing have no way

to prevent GM crops cross-pollinating with traditionally bred crops. So, an organic farm growing the same crop might find that the manipulated genes end up in their crop. They do not have any choice in this matter, so feel that they are being invaded by unwanted materials. But the word 'contaminate' is curious. It used to apply to 'foreign bodies' in otherwise clean environments. It usually applies to chemicals polluting the air and water. To use this word to describe the same biochemical substances turning up amongst the same molecules is curious. Perhaps there should be a new word. But mixing genes is what makes the world go round.

6. Many people do not like the idea of picking out a virtuous trait in one creature and putting it into something completely different. The idea of a fish gene in a tomato is particularly gut wrenching. This is transgenic GM. There are pushes to stick to cisgenic GM – i.e. genes between the same species. It does feel odd, until you realise that most creatures in the world have the same set of four chemicals determining who and what we are. Our set of genes is 99 per cent the same as monkeys. Even fish have the same chemicals. Eric Morecambe could have said: 'they are the same genes, but not necessarily in the right order.'

7. There is concern that GM increases the use of pesticides. The big corporates have so far developed two main sorts of GM. Monsanto's 'Roundup Ready' increases the use of herbicides or weed-killers because it enables corn to grow while all other weeds are destroyed. The Roundup refers to glyphosate, as we have seen. There are new moves to bring in a GM variety that makes crops tolerant to 2,4-D herbicide, another widely used weed-killer. 2,4-D has been around a long time with little evidence of toxicity affecting the environment.

The other main GM is 'Bt' which produces an internal natural insecticide. It means farmers use less insecticide on crops like cotton and maize, which is the best way to protect workers in developing countries. There is some concern that the Bt may affect beneficial insects and kill off those that act as biological control agents. The Organisation for Economic Cooperation and Development (OECD) looked into this and found no case.[40] Use in China indicated that while Bt cotton dealt with bollworm, other pests then moved in. That would seem like a quite natural thing to happen.

The biggest concerns now are that many weeds in GM crops are growing resistant to the herbicides. Prince Charles picked up on this in 1999, saying 'there is already evidence that the genes for herbicide resistance can spread to wild relatives of crop plants, leaving us with weeds resistant to weed-killer.[41]

That is almost inevitable. Pests and diseases throughout history have built up resistance to what we've thrown at them. That is why they are pests. It is part of the never-ending war with the bits of nature we don't like and is not confined to GM. There are millions of acres in the US where weeds that were being killed have built up resistance to glyphosate.[42] In India the Bt cotton control of bollworm has broken down.[43] The answer by GM companies is to 'stack' the genes so that there is a multiplicity of them, making it 'virtually impossible' to develop resistance to them all. We shall see.

8. Some, like me, say GM encourages a 'reductionist' science. 'Reductionist' science is where a massive amount of complex, interacting data is 'reduced' to a single number or unit.

By reducing the problem of producing more food, more sustainably and more healthily to a few genes, this misses much of the analysis about society and hence possible solutions. A lot of complex global-warming issues are reduced to carbon-counting. Much of science is reductionist, not just GM, as many scientists do not want to see beyond their microscopes.

The best/worst example of this is where corporate advertising claims that GM will 'save the world from hunger'. Of course it won't. In the opening chapter, I showed how hunger in the world has little to do with how much food is produced. It has a lot more to do with whether people can afford to be in the market economy. GM won't solve that. But don't blame GM for making that claim. There is no technological fix, nor any one solution. But we should have GM in our toolkit.

9. I remember a report in *Nature* purporting to show that GM killed Monarch butterflies – it shocked me.[44] A lab test concluded that pollen from Bt corn (with the internal insecticide) that killed pests could also kill Monarch butterfly caterpillars.[45] To lepidopterists like me this was the end. But field trials showed it wasn't true, again showing the importance of field trials compared to lab work.

10. Some consider GM a 'sin of science'. Peter Melchett, in his article 'Seven Sins of Science',[46] said the first sin was: 'Pro-GM scientists [who] have made the mistake of conflating their opponents' opposition to commercial products (GM crops) with opposition to science'. He goes on: 'they have consistently promoted GM in ways which are not only unscientific, but which have been positively damaging to the integrity

of science.' This brings us back to the 'March of Science' in the last chapter; avoid making such accusations, but look at the evidence. Again and again.[47]

CITIZEN'S PANEL

Plant scientists have recommended a new programme of independent research to field test 'public good' GM crops. The proposal, which they are calling PubGM,[48] would support innovations to benefit consumers and the environment. This comes in a report to the Council for Science and Technology,[49] the body that advises the Prime Minister on science policy issues. Leading plant scientists from the universities of Cambridge and Reading, the Sainsbury Laboratory and Rothamsted Research also recommend that legislation for GM crops should be decided on a national level, in a similar way to pharmaceuticals.

As a result of Brexit, there is an opportunity now to have a proper – dare I say scientific? – debate. 'Crop varieties improved by GM methods can help maintain or increase harvests while reducing environmental impact, but we need to assess their usefulness out in the field', says Professor Jonathan Jones from the Sainsbury Laboratory. 'With PubGM, seed companies, consumers and regulators will be able to decide, based on results of experiments, whether a GM trait has proved its worth in UK crops under UK conditions.'[50]

Retailers are still shy of taking it on, but they need to be involved. In 2010, they went to the government to get approval for selling chickens fed with GM corn, without having to label them as containing GM.[51] They seem to think everybody is bothered, yet when the people were asked in the UK about their food concerns in 2014, GM came seventh behind food hygiene, food poisoning, labelling, additives, hormones and pesticides.[52]

While I was on the Co-op's Responsible Retail Committee, we were asked for advice about introducing biodegradable, instead of oil-based, plastic packing for their sandwiches. However, the only source of a million such sandwich cartons a day was Cargill, who reminded us that the corn used to make the biodegradable plastic would be GM corn. There was a close vote to say they shouldn't go ahead with that. I thought they should, as environmentally there are many more impacts arising from oil-based plastic than biodegradable corn plastic, whether or not GM.

GM isn't a perfect technology, but what is? I have found only one fatality due to GM. A farmer in America in 2015 was killed when a tonne bag of GM corn fell on him. This is not a sick joke, but a reminder of where the real killers are – old-fashioned accidents in dangerous workplaces.

Whatever is wrong with the food and farm system in Britain – and I have spelt out quite a lot in this book – it has nothing to do with GM, because there isn't any here. Yet there are too many farm fatalities, awful working conditions, insufficient home food production, rich landowners, degraded soils, too much nitrate fertiliser, too much flooding, poor dairy farmers, obesity, food banks, food poverty and colossal food waste. Please do not use anti-GM as a proxy for all our food and farming ills.

PART IV

The Future

11
Favourite Foods

This chapter looks at how some of our favourite foods might be affected by the Brexit process. Let's start the day with our famous breakfast.

Full British 'Brexfast'

Eggs

At present, importing eggs from outside the EU is subject to a 'third country' duty of E30.40 per 100kg. We look up these values as commodity codes; this is 0407210000 for eggs, in shells, fresh or processed of fowls of the species *Gallus domesticus*.[1] When we leave the Customs Union, our eggs will be subject to this tariff when exported to the EU. We will be considered a 'third country' unless, when negotiating a Free Trade Agreement (FTA) with the EU, a tariff-free arrangement is made – as we have now. Retailers are turning away from expensive European eggs – because the pound has slumped. The man in charge of the Lion Egg Processors expects this 'significant rise to continue because of fears of tariffs in the long term'.

Bacon

As part of Brexit, it would be wonderful, in the great untangling of EU directives and regulations, to loose Directive 2001/89/EC.[2] This controls swine fever by banning the use of food waste for pig swill. The EU prohibited the use of 'pig swill' – despite this process having been used for centuries, all around the world. The authorities blamed poorly-cooked pig swill for the Foot and Mouth outbreak in the UK in 2001 that destroyed so many cows and sheep. We should be able to transform food waste to swill safely.

Brexit is blamed for the increased price of bacon, rising by 40 per cent from July to August 2016 after the Referendum, fuelled by the slump in the value of the pound. There was also a surge in demand for British pork exports, mainly by China, due to flooding in one of its large pork-rearing areas. By early 2017 prices were down and sales increased.[3] This demonstrates the volatility of pig prices – we were taught in agricultural economics that they display a famous 2-year cycle.

Can we buy more British bacon? Bacon is a £1.3bn industry in the UK.[4] Yet 75 per cent of what we buy comes from the EU.[5] In 2001, Japan introduced the Promotion and Utilisation of Recycled Food Waste Act, which promotes pigs reared on food swill as an eco-food. Our costs could be reduced and the pig-swill-reared product sold as 'eco-friendly'.

Mushrooms

Two-thirds of all Britain's imported mushrooms come from Ireland, most of which are from Monahan mushrooms. Irish mushroom farmers were the first to be hit by Brexit and the pound's loss of value. Out of 60 mushroom companies, 5 had gone out of business by November 2016. Irish mushroom growers rely on the UK for 80 per cent of their sales and have been losing money since the vote in June 2016. Ireland's mushroom industry, the fifth largest in the EU, has an annual production worth €120m (£108m) and employs about 3,500 workers. They – and other agri-food suppliers who are also suffering – were offered cheaper loans and modest tax relief. The head of the Irish Creamery Milk Suppliers Association likened these as 'giving someone a bicycle when they needed something with a motorised engine'.

Cereal Killers

Cereals took over from the traditional British breakfast in the 1970s and 1980s. The most famous cornflake manufacturer is Kellogg's, who have always claimed to be healthy. 'Kellogg was the first company to print nutrition messages and product information on cereal boxes so that we all could make more informed decisions about the foods we eat.'[6] While it could just be a coincidence, the rise in UK obesity parallels the increase in breakfast cereal consumption.

The first new Kellogg's advert in five years featured two sets of Brummie twins[7] (I'm a Brummie twin) in January 2017. It didn't mention the Public Health England report[8] that had appeared 3 days earlier, which said the average child is eating 3 cubes of sugar every morning for breakfast – half their maximum allowance – before they go to school. 'It found that of those parents whose child was consuming the equivalent of three or more sugar cubes in their breakfast, 84 per cent considered their child's breakfast as healthy.' In July 2016, Kellogg's were censured by UK Authority for claiming that their Special K porridge was 'full of goodness', as this contravened rules laid down by … the EU.[9]

Credit Suisse business advisers say that because Kellogg's sell 20 per cent of their produce to Europe, they are very exposed. They will have to cope with the probability of a recession in the UK, the immediate prospects of higher raw-material costs and higher tariffs across Europe.

The issue about sugar on cornflakes goes right back to the founding Kellogg's brothers. John Kellogg is generally credited with creating a whole-grain cornflake that was considered healthier than the meat/egg breakfast of the wealthy, and the porridge gruel of the poor. He ran a sanatorium, using holistic methods particularly in relation to nutrition, and was a vegetarian who believed the cereal would both improve the nation's health, and would keep people from masturbating and desiring sex. However his brother, W.K. Kellogg, was more interested in the business side, so he added sugar to the cereals, much to his brother's annoyance. Kellogg's expanded by 25 per cent in 5 years in the 1970s. All of the corn used for UK cornflakes comes up the Manchester Ship Canal from America. If the value of the pound stays low, it will inevitably cost more. We import around $500m worth of corn.

While fatty breakfasts were blamed for making us fat, the bitter truth is that sugar is much more culpable. When we have control over our borders, we could put limits on the amount of sugars and HFCS in breakfast cereals, to stop the invasion of cheap sugary foods.

Sugars

'Doing a deal' will involve sugars. The EU call High Fructose Corn Syrup (HFCS) 'isoglucose'. In biochemistry we learnt that fructose and glucose are isomers of each other, hence the confusion of names. Isoglucose (i.e. HFCS) carries the largest EU tariff of 600 per cent. We saw the possible health impacts of fructose in the chapter on obesity, but the tariff isn't there to protect our health; it is to protect the health of the sugar industry. It undermines the price of other sugars, because it is the cheapest sweetener.

Sugar tariffs and associated quotas were to protect the EU sugar industry, particularly sugar beet in France, developed originally by Napoleon. However a ruling from the World Trade Organisation said the tariffs had to go – they were a barrier to free trade. So a major EU sugar reform was carried out in 2006, allowing more tariff-free sugar to come into the EU.[10]

The EU also put a stop to the sugar industry getting massive subsidies as 'non-farm' payments. These paid sugar manufacturers to export the stuff (dumping). In the two years 2004–5, the UK sugar industry received over £250,000 to fund their exports of sugar and isoglucose.[11]

There will be no internal quotas on growing sugar within the EU after 2017. The amount of isoglucose produced (from maize) is expected to increase from 700,000 tonnes to 2.3m tonnes. American breakfast cereals are drenched with isoglucose.

As we will 'control our borders', we will be able to change the sugar tariffs and quotas. WTO rules allow us to have phytosanitary (SPS) barriers – i.e. protect our health, safety and the environment – but only to the same standards we impose internally. I propose we put the same 'sugar tax' we are putting on sugary drinks onto sugar imports. But that leaves the way for tariff-free isoglucose.

The vote to leave the EU has sparked a clash between sugar-cane importers and farmers producing home-grown sugar beet. This bittersweet skirmish is a microcosm of the bigger food/farm battles to come. Every time we drop import tariffs, cheaper food will come in, threatening farming livelihoods here. Sugar beet (produced in East England) reflects the earthy side of Britain's character, while sugar cane reminds us of our colonial roots and the darker history of Britain's part in the slave trade.

Tate & Lyle claim they can save another £20 million if they could have free trade agreements.[12] Tate & Lyle actively campaigned for Brexit and always moaned about protecting 'French beet farmers'. In their exultation over free trade, the Brexiteers' promise of 'taking back control' means here favouring foreign importers and owners, American Sugar Refining, who bought Tate & Lyle in 2011.

Sugar beet farmers are realising what could be hitting them. Sugar beet is a reliable source of income for 3,500 of Britain's arable farmers, supporting 9,600 jobs. Sugar beet's broad green leaves act as a rotational crop between soil-sapping wheat. Beet produces 'Silver Spoon' the sugar brand of British Sugar, owned by Associated British Foods. The NFU fear that the economics of this industry will unravel.

We could use a lot more locally-grown sugar beet and maize to fuel cars. About a quarter of all maize grown in the USA is converted to bioethanol used to fuel cars. Much Brazilian sugar cane now goes into fuel. British Sugar set up a biofuel plant based on sugar beet to make bioethanol – a 'cleaner fuel' – at Wissington in Norfolk in 2007.[13]

Toast and Jam

While there are all sorts of breads, most are made from wheat. The EU takes the lion's share of our grain exports – 70 per cent. Our future grain production and prices are at the mercy of powerful trading forces in the world, particularly the prospect of increased Russian grain production. Four consecutive years of global grain surpluses mean world grain prices have reached a low point – another example of the mad food markets at work. Only currency devaluation insulates the UK market.

If we are thinking of going global, an AHDB Report for Grains in 2016 says: 'The balance of the world's top wheat exporters is changing. Since the Great Grain Robbery in 1972 (when US Secretary of Agriculture, Earl Butz sold grain to Russia in secret, and was heard to say "food is a weapon"), the share of global wheat exports attributed to Russia has increased from 2 per cent in 1972/73 to a projected 31 per cent in 2016/17 ... In 2016/17 Russia is forecast by the International Grains Council (IGC) to become the largest global wheat exporter, with 30.7mt entering the world market.'[14] We cannot compete with that and go global.

The same report says, of the prospects for buying British: 'The UK is set for interesting times with the availability of both wheat and barley feeling the squeeze on the back of reduced production levels. The AHDB Cereals & Oilseeds Early Balance Sheets for Wheat and Barley October 2016, forecast a 12 per cent fall in UK wheat production in 2016/17.'[15]

About 85 per cent of the wheat used by UK flour millers is home-grown. We import just over 10 per cent, mainly from Canada, Germany and France. Canadian wheat has stronger gluten strength, so is mixed with British wheat to make bread dough stronger and more elastic.

For spreads on your toast, Liam Fox, Secretary for International Trade, tweeted that France needs innovative British jams and marmalades. Somebody should tell him they already have classy jams – called Bonne Maman.

But when is a jam a jam? A war erupted a few years ago,[16] when some UK manufacturers wanted to call their jams with reduced sugar content 'jam', but were told anything below 60 per cent sugar had to be called 'conserve' or 'fruit spread'. They were told that the EU Jam and

Similar Products Regulations 2003 prevented them from calling it jam. Newspaper headlines screamed 'EU Red tape', 'ridiculous EU jam laws' and 'the end of the British breakfast as we know it'. It eventually emerged that member states could make exemptions to the rule for their own traditional products.[17] It was just that DEFRA didn't seem to know. The irony is that the particular EU law in question came from *CODEX Alimentarius* – the world body to which we would be subject after Brexit.

Coffee

When you order Cappuccino – the name based on the colour of monks' cloaks – you become part of the second biggest traded commodity in the world. We import nearly £1bn worth of coffee.[18] There are no tariffs on raw beans but the EU imposes a 7.5 per cent charge on roasted coffee.[19] Coffee trading shows us the complexity of the foodstuff tariffs that the EU imposes.

The TARIFFS DATABASE, called *TARIC*,[20] is where you can find out the tax on any of 2,000 agricultural goods. You need the 'goods classification code' from the government site 'Trade Tariff Commodity Code'. On leaving the EU we will be classed, by default, as a third country, quoted on the government site, so we can see what tariffs we are likely to attract on leaving the EU Customs Union.[21]

Choose 'Vegetable Products' (Section 2), which leads to 'coffee, tea, mate and spices' (09). Choose 'coffee' (01) and then you can choose 'coffee not roasted'. The codes open up on the right-hand side for caffeinated and decaffeinated.[22] Table 7 shows the various combinations. They are not the same.

Table 7 Coffee Tariffs

Coffee Not Roasted	Chapter 9	Third country	Ghana
Not Decaffeinated	0901 11 ▼	0 per cent	0 per cent
Decaffeinated	0901 12 ▼	8.3 per cent	0 per cent
Coffee Roasted			
Not Decaffeinated	0901 21 ▼	7.5 per cent	0 per cent
Decaffeinated	0901 22 ▼	9.0 per cent	0 per cent

This is about protecting our coffee-roasters. There isn't a tariff on *unroasted* coffee, so exporters from Africa and South America export unroasted coffee. It is not worth their while investing in roasting/decaffeinating technology. There are special deals, like Ghana, which are still protected against all tariffs. The people who really make the money out of our old colonial friends are European roasters – protected by the particular way the tax is imposed.

We export $318m worth of coffee, despite not growing a bean; over $50m to France and around $35m to Ireland.[23] At the moment there are no tariffs involved with these exports, but this would change outside the Customs Union. The default 'third country' position is that when we export roasted coffee to the EU post-Brexit, we will pay 7.5 per cent on the sale.

When we go global, we could help our old colonial friends by removing the roast tax and encouraging the capital development of coffee cultures in the host countries, instead of leaving the producers exposed to volatile world market prices.

Lunch

Sandwiches

The British Sandwich and Food to Go Association (BSA) is getting 'excited' about Brexit. It opens up opportunities to ban some EU food labelling regulations. In particular, they could change 'use by' to 'use by end of' on sandwich packaging. They believe it could turn the tide on food waste. The Association sent a letter to Brexit secretary David Davis, saying: 'We would welcome the flexibility to return to this as we believe it is both clearer to consumers and to retailers. This simple change would have a significant impact on food wastage at the end of shelf-life.' They say this wording shows that EU legislators misunderstood the UK sandwich market when they introduced the 'Food Information Regulation'. Apparently 'the UK leads by some way as a European nation in the food to go sector'. Other EU markets are less developed, so this 'one size fits all' law does not fit us. I can't see this is a great reason for coming out of the largest Single Market in the world. Does it mean we can leave our sandwiches to get soggy for an extra day?

The sandwich and 'food to go' sectors are heavily reliant on manual labour – much more than other food sectors. Most of this is migrant labour. If a 'points based' system is introduced, it would penalise unskilled labour further in this sector. The BSA think they can get round this by introducing limited working periods, provided 'training' is included in each.[24] Pret a Manger manager Andrea Wareham is worried about filling jobs following Brexit, saying she will struggle if the company has to hire only British workers. At a House of Lords Economic Affairs Committee, in March 2017, Pret said just 1 in 50 people applying for jobs at the chain were British. The committee suggested hiking pay above a starting salary of £16,000 might help. Wareham said Pret staff can increase their salary to £45,000 including bonuses 'within a few years', adding: 'I actually don't think increasing pay would do the trick, we have great benefits and we offer fantastic careers and the company is going on a recruitment drive to hire British workers.'[25]

Pizza

The then PM David Cameron decided to hold the referendum on this country's membership of the European Union whilst in a pizza restaurant at Chicago's O'Hare airport.[26] An average Brit spends £4,000 a year on dining out, twice the amount in the 1970s. The restaurant trade in 2017 is big business, after being knocked sideways by the banking crisis in 2008. Eating out was the first thing people stopped doing. The reliance on EU staff in restaurants has increased by nearly 50 per cent over recent years. Over 1.3m staff would need to be recruited into the sector between 2014 and 2024 to keep pace with demand. The stronger euro compared to the pound means these workers have less money to send home. In real terms they have experienced a 20 per cent pay cut since the Brexit vote. In March 2017, a restaurant in Brighton, part of an Italian chain, had a third of the tables sectioned off, plus a reduced menu, due to staff shortages.

The Living Wage has risen to £9/hr, and is welcomed by most of us. However, it does mean that if restaurants want to attract low skilled workers from other sectors, they will have to pay more. Together with the higher costs of ingredients, expected to increase costs by another 3.4 per cent in 2017, accountancy firm Moor Stephens has estimated that over 5,500 restaurant companies will go out of business in the next three years.[27]

Pre-packaged Meals

Soybeans can be made into flour, soymilk, tofu, oil and a cheap protein filler found in many pre-packed meals. Imported as beans, they do not attract any EU tariff. Soya oil (commodity code 1507109000) attracts a 6.4 per cent tariff, again protecting our soya processers against the primary producers.

It is hard to know how we can go global with soya when we import \$360m worth of soya beans, 60 per cent from Brazil and 30 per cent from the USA. We also import £880m worth of soybean meal (9 per cent tariff) mainly to feed cattle. Once outside the EU, this 9 per cent tariff will be imposed on the prepared soya-based foodstuffs (PAPs) that we export to the biggest Single Market in the world. In the prepared foodstuffs segment, we should look to soya substitutes, such as rapeseed meal and extruded peas, both of which are easily grown in the UK.

Burgers

We will have less chance of knowing what is in our burger.[28] In January 2013, the UK Food Standards Agency (FSA) was notified by the Food Safety Authority of Ireland that its survey of processed beef products had identified trace amounts of horse DNA in the majority of the samples. It identified one product, a Tesco burger, as flagrantly adulterated with horsemeat.[29] I wasn't surprised when an abattoir near Todmorden was investigated as a source of the contamination, as the Pennine hills have loads of nags. We could put a stop to that following Brexit, and pay farmers only for growing 'biomass' – willow coppicing works well on this sort of land.

Food fraud goes on particularly when the price of ingredients goes up – as they are now. Prior to the Brexit vote, 60 per cent of UK food manufacturers believed that leaving the EU would be bad for their businesses – because of possible pressures on prices. It is likely that meat ingredients will start being sourced from outside the EU, particularly cheap US beef – or 'hormone beef' as we like to call it. While the EU bans it, we won't have to. But if we want to sell our meat products into the EU, we will have to exclude lower standard meat. We may start seeing two sorts of meat-processing factories, working to different food standards.

Afternoon Tea

Tea

Our favourite fix is 'A nice cuppa'. This term conjures up a picture of innocence while hiding some of the worst atrocities in world food history. The classic British cuppa hides a history that shaped much of how the world is now. We stole the tea plants that we now grow in India and East Africa from China. A Scot called Robert Fortune shipped plants to grow in India, although only 80 out of 13,000 plants survived. All Indian and East African plants are based on this limited range of plants. We would now call this industrial espionage, and it served to break the Chinese monopoly on tea. To feed our tea addiction, we doped the Chinese with opium. The East India Company grew 100,000 acres of it and sold it to middlemen who sold it in China to get silver and gold to buy tea. That company makes Monsanto look like a choir of angels. When the Chinese burnt a cargo of opium in the harbour, we sent in the Navy to put them down. They were made to open up their ports to 'free trade' and had to give us Hong Kong. Our colonial exploits may yet come back to bite us. Tea also featured in the American War of Independence. The Boston Tea Party occurred when demonstrators against a tea tax destroyed an entire shipment of tea sent by the British East India Company. The British government supported the company by helping them offload £17m worth of excess tea.

When it comes to Brexit, tea is probably the least contentious issue. The EU Food Standards have some impact on tea imports. Researchers found that pesticide limits in food applied by importing countries significantly reduced tea exports from China.[30] Chapter 9 noted the standards called Maximum Residue Levels (MRLs), which indicate how much pesticide is allowed in a foodstuff.

It means tea from China has trouble getting into the EU – despite it being the originator of the crop. We could drop the standards to enable cheap Chinese tea in. The UK could use its new-found 'risk assessment' rather than 'hazards' approach to let the tea in. The MRLs for tea are based on the hazards approach – a set cut-off point. A risk-assessed approach could let it in. But the risk to other sales could put them off.

We import $372m worth of tea. There are no EU tariffs on black tea (0902300000 commodity code attracts zero rate[31]), although there is a small tax on green tea. That is because there are no (well very few[32]) tea-growers to protect in the EU, nor a tea-processing industry. When it comes to doing a deal, while we've seen how tax and tea can be a pretty lethal combination, I think it unlikely there will be any tariff changes.

We could go global as we already export $148k of tea. If you were to ask how is it that we can export that amount when we grow nothing like it, apparently 'We don't have to grow tea to turn it into a distinctly British product any more than we grow the oranges that make British marmalade. What matters isn't where the raw materials come from. What matters is the value you add', according to the free trading *Telegraph.*[33]

Milk

After Brexit, a minister promised to keep the EU's £10m 'subsidy' of school milk.[34] NFU representatives met DEFRA officials to discuss the future of school milk in the UK, which has participated in the EU scheme. Children over the age of five receive a subsidised portion of milk every day in school. Michael Oakes, the NFU dairy board chair, said: 'Given the importance of school milk for our children's nutrition and shaping future consumption trends, we were pleased with the outcomes of the meeting we had.'[35] The EU scheme was criticised for being costly to administer and the European Dairy Association has described its functioning as 'suboptimal in most member states'. DEFRA ministers are open to suggestions for what a new domestic scheme could look like and plan to consult on the idea in due course.

When doing any deal, we have to consider that 80 per cent of our dairy exports are to the EU and 96 per cent of our dairy imports are from the EU. Much of that is in the form of cheese. Nevertheless, once outside the Customs Union, there will be a 30 per cent tariff on any movement of dairy products across the border.

The prospects for going global don't look good. The trade in large quantities of UK milk to Ireland (mainly from Northern Ireland) for cheese-processing would face a Tariff Cliff, which varies, but milk with over 3 per cent fat faces a tariff of nearly 20 euros per 100g.[36] There is nowhere else worth exporting to.

Buying British should be a priority. Dairy farms are already hard pressed. At present, because of overproduction, the costs of production often exceed retail prices. To deal with the overproduction of milk in the UK, the most obvious solution is to drink more. There should be a campaign to replace fizzy sweet drinks with milk after strenuous exercise. Dr John Ivy compared the recovery benefits of drinking low-fat chocolate milk after exercise to the effects of a carbohydrate beverage with the same ingredients and calories as typical sports drinks as well as to a calorie-free beverage. Ten trained cyclists had significantly more power and rode faster when they consumed the low-fat chocolate milk, and had twice the improvement in maximal oxygen uptake, a good indicator of aerobic ability, after a month.[37]

Ice Cream

On Brexit, ice cream labels may change, since many of the labels are based on EU standards that were opposed by the British. Our ice cream industry argued that we do not expect cream in ice cream, quoting cream crackers as an example. Under UK law, enshrined in the Food Labelling Regulations 1996, products called 'ice cream' must have at least 5 per cent fat and 2.5 per cent milk protein. In the rest of the EU they say 'ice cream must contain fat (from dairy, eggs or vegetables)', but do not set a minimum fat or protein level. The Food Standards Agency said we should align with Europe. Unilever agreed, saying alignment was important for global business.

In 2016, the Unilever boss warned that the price of Magnum would rise on Brexit because we would have to pay 30–40 per cent import duties on milk, sending ripples of fear across the nation.[38] We don't buy British milk/cream to make Magnums, because while we eat 100 million of these 'pinnacles of indulgence', they don't contain cream. Instead they contain corn and rapeseed oil. A speaker for Unilever said: 'It's in line with our policy of removing saturated fats.'[39] Other brand label ice creams use coconut and palm oils.

Cakes and Biscuits

When doing a deal, we find cakes and biscuits are called 'Processed Agricultural Products' (PAPs). If 2,000 tariffs are complicated enough, then they take on a new dimension when it comes to cakes and biscuits, because these are made up of a mixture of farm products, like fats, milk and sugars, each attracting different rates of tariffs. It means there are 15,000 – yes, 15,000 – different PAPs with their own quotas and tariffs. The recently signed FTA between the EU and Canada (CETA) took years to negotiate and was delayed in its final stages by a regional parliament in Belgium.[40] This deal does away with tariffs on PAPs, but it depends whether we Brexit before or after that deal is ratified as to whether we are part of it. Before, we are not; after, we are bound – although it is not quite that straightforward.[41]

As we go global selling tea and biscuits in our great export plan, it is predicted biscuits will bring in more than £300m. One of our most famous is McVities Digestive and its makers, United Biscuits, reckon export sales rose by half from 2012 to 2015, shipping 60,000 tonnes to more than 100 countries. However, the owners of this quintessential

British biscuit are Turkish. It is they who are planning a major push to sell Digestives, Jaffa Cakes and Hobnobs to Americans.

At the Bar

Beer

It is nearly impossible to say what may happen with the messy legal tangle of regulations that govern the beer trade within the EU. A brewery like BrewDog, which owns dozens of bars across Europe from its headquarters in Scotland, may have to re-negotiate contracts with landlords, labour boards, suppliers and alcohol control boards in every one of those countries.

Currently, the UK imports ($35m) about twice the amount it exports ($18m).

When going global, beer features in the government's International Action Plan for Food,[42] as we saw in Chapter 4, and is expected to gain £300m worth in exports. They also expect to export £20m/yr of barley to fuel the fast-growing beer market in China. Forbes in America says: 'English imports like Sam Smith or Old Speckled Hen might give American craft beers a jog for their money.'[43]

It will become very difficult for British breweries to grow, even survive, in the medium term. 'Brexit means a lot of uncertainty for our business', East London Brewing Company director Claire Ashbridge-Thomlinson told the Good Beer Hunting blog.[44] 'As a concrete example of this, we have just had a cancellation of an export order from an Italian customer who felt insecure going ahead in the present climate.'

It's not bad news for *all* UK beverage companies. London-based Diageo, owners of Guinness and many of the world's beverage brands, is expected to fare well. With less than a reported 10 per cent of its product staying in the UK, Diageo should benefit from trade with countries around the globe who can purchase British booze cheaper than before. After the Brexit vote, Goldman Sachs upgraded Diageo from 'sell' to 'hold', giving it leverage against a buyout – something many analysts predict will happen.

We are the third largest importer of hops in the world, importing $35m with over $14m of our imports from the US and $10m from each of Germany and France.[45] Curiously the UK exports nearly $6m hops back to the US (different varieties) and $2.5m to Germany. Hops imported from third countries attract a tariff of 5.4 per cent.

There used to be hop fields all around the English-Welsh borders: I worked on a Guinness hop farm there. Now there are far fewer hop fields. Hops were considered so important before the Second World War that there was a Hop Marketing Board which set up quotas, guaranteeing decent prices. I remember the Hops Research Station at Wye, now long gone.

British hops share the same 'terroir' (meaning 'earth', or character through the growing) needing no irrigation, so are more sustainable than hops grown elsewhere and produce a better aroma as they have lower 'myrcene' levels, a component of hop oils that gives the beer a peppery and balsam smell.

The newer 'craft' beers focus on an intensity of flavour – the fruitiness of the hops. UK hops are responding, so there are now many UK craft beers, driving the industry. It is a fabulous example of how we could build up our food systems based on selling more to ourselves. It's a shame we haven't still got that Hop Research Station at Wye.

Whisky

'Hard liquor' – whisky and gin – accounts for 40 per cent of all our food and drink exports, some $8bn.[46] Whisky is the most important, with more than 90 per cent of Scotch whisky sold outside the UK, making it the biggest single net contributor to the UK's balance of trade in food and drinks. A fifth of the exports are to the USA and almost a third to the EU.

The Scotch Whisky Association (SWA) is confident that certain things won't change, as there is no EU tariff. They do not expect any changes from India, who impose a massive 150 per cent tax. They believe Brexit won't make matters any worse and there may a chance for bilateral deals. They suggest the prospect of perhaps doing a deal with India in the long run by offering help with improving water quality in India – something the whisky industry is good at. But they know some things will change, like 'rules of origin' making administrative arrangements crossing borders more difficult. Unless there are transitional arrangements, Scotch 'will lose significant tariff reductions in certain markets, notably Korea, South Africa, and Colombia and Peru'. They say that if the UK needs to rely on WTO rules, 'this will take a major upgrade of capacity within the UK Government and can't be done quickly'.[47]

The whisky industry wants an open and free-trade policy – the sort we are walking away from – and agreement with the EU on practical arrangements. They want clarity from the government on customs enforcement, and a close working relationship between Edinburgh and London with 'a seat at the table' when discussion takes place about all the rules and regulations that may change – including the very definition of 'whisky'.

Wines

Following the Referendum result, wines from Europe are becoming more expensive. The country's leading wine trade body, the Wine and Spirit Trade Association (WSTA), says: 'Consumers may soon have to pay an extra 29 pence on average for a bottle of wine from the European Union if cost increases are passed on, due to the devaluation of the pound. There also is the prospect of tariffs on French, Italian, Spanish and other EU wines if the UK leaves the single market.'[48]

We could buy more British wine to reduce that $4.3bn import bill. We are growing wine grapes as far north as Morecambe. England has 133 wineries, which produced 5m bottles in 2015. However, our crops can get hit by frosts like those in April 2017. Half the crop in the UK could have been lost to 'air frost'. Growers can cope with ground frost using hundreds of burners, but not air frost. If ever there was a case for a British Wine Research station, this is it.

In 2016, home-grown British bubbly scooped seven top international awards.[49] Some people are saying some of the UK sparkling wines are better than champagne.[50] The European Union is worried that 'British champagne' could flood the continent after the UK leaves the EU.[51] Under current EU law, more than 1,000 food and drink products, 59 of which are British, have a 'Geographical Indication Status' (GI). This means that they can only be produced in officially-recognised areas. Champagne can only be produced in the Champagne region of France. It seems that the UK could start selling 'British champagne'. The European Parliament's Agriculture Committee drafted a warning: 'As things currently stand, the UK has 59 such registered names [out of 1,150 at EU level], including e.g. Lakeland Herdwick Meat, West Country Farmhouse Cheddar Cheese, West Wales Coracle Caught Sewin [sea trout], and [economically important] Scotch Whisky.' They say that it is important that in any deal, there be mutual recognition of GIs.[52]

Evening Meals

Beef

Most beef imports into the EU are subject to tariffs, some 50 per cent. There are tariffs on processed beef products, and quotas (around 120,000 tonnes) for reduced tariffs on high-quality beef from just about everywhere. If no trade deal has been agreed with the EU when the UK leaves, UK beef would be subject to these tariffs when entering the EU. The trade could involve payment of two tariffs, probably making it uneconomic for both dairy and beef.

When going global, at present, there are no significant EU Free Trade Agreements involving beef, but the EU are negotiating with several major beef exporters, from which we will be excluded. The main issue is the Sanitary and Phytosanitary (SPS) barriers – beef treated with growth hormones. This will a big issue following Brexit, and particularly in any US-UK trade deal. The US are clearly eyeing up an opening,[53] despite reassurances from the Agricultural Minister that standards won't be lowered.[54]

'Hormone beef' has been a long-running argument between the US and the EU. In the US, they routinely inject a variety of six synthetic hormones that encourage growth, so that a $3 dollar injection produces extra meat worth $30–40. In 1981, following consumer pressures especially from Italy, the EU prohibited the use of substances having an hormonal action for growth promotion in farm animals. These were synthetic versions of oestradiol 17ß, testosterone, progesterone, zeranol, trenbolone acetate and melengestrol acetate (MGA).[55] In 1989 they went on to impose this ban on imported beef. The US retaliated and imposed tariffs on selected food goods. The list of 'retaliated' products does not include any from the UK, although Roquefort cheese is clearly aimed to hit France. They took the EU to the WTO in a long-running battle. In 1998 the WTO ruled in favour of the US, citing lack of 'risk assessment' in the EU, so the US upped the tariffs. The EU dropped the ban on most of the hormones, leaving only oestradiol banned. In 2009 the US 'escalated' the tariffs, and eventually in 2012 the EU and the US came to a deal, allowing a quota of US imports of high-quality beef.[56] But they still feel their export markets have been restricted, and it will be interesting to see how President Trump supports his family farmers in any US-UK trade deal.

In terms of exporting beef, we still don't have access to many countries, as trade has not been restored following the BSE-related restriction in the late 1990s. Exports are likely to be limited to premium cuts and lower-value cuts and offals.[57]

We consume over a million tonnes of beef and veal, of which about two-thirds are produced in this country.[58] The prospect of buying more British beef – up to around 400,000 tonnes – would seem easier than going global. We will have difficulty if tariffs on beef coming in are removed and we are swamped by hormone-drenched beef.

Chicken

Following Edwina Curry's gaffe in the late 1980s about *Salmonella* being 'endemic in our flock', in the UK we have introduced measures to reduce dramatically the infection throughout the rearing process. The rest of Europe shares our preventative approach. But in America and elsewhere, they get rid of the bug by washing poultry with chlorine dioxide or acidified sodium chlorite. It means chicken infected with salmonella can enter the food chain. This 'easy fix for dirty meat' reduces incentives for farmers to treat infections early on.

If we try to go global and make a free trade agreement with the US, the US will want to sell chicken to us. There will be an outcry about their chlorinated chickens. We could ensure SPS measures, as we maintain those standards ourselves, but we could still expect a US challenge to the scientific justification.

When arranging a deal about chickens, the commodity code for importing from outside the EU is 0207111000 (83 per cent of the bits of *Gallus domesticus*), for which the tariff is 26.20 euros/100 kg. Other chicken bits in different forms (e.g. frozen is 29 per cent) are much the same.

We import $1.6bn worth of poultry meat, nearly half from the Netherlands and $245m from Poland, and around $100m from Germany, all of which comes in without tariffs. I was surprised that 'only' $12m comes from Thailand. We export $417m, $129m of which goes to Ireland.

We should look to produce more high-quality chickens for the UK market. In the process, the labour to do it should be better rewarded.

Pork

On Brexit, we could lose the Council Directive 2008/120/EC[59] which lays down minimum standards for pig protection including: improving the quality of the flooring surfaces, increasing the living space available for sows and gilts, higher levels of training and competence on welfare issues for personnel, minimum requirements for light and maximum noise levels, and access to fresh water and to materials for rooting and playing.

Since January 2013, pregnant sows are kept in groups instead of individual stalls during part of their pregnancy – a major improvement for the welfare of sows in the EU. With some exceptions, all pigs are to be raised in groups and must be provided with a sufficient quantity of enrichment materials, so pig farmers carry out proper investigation and fulfil their behavioural needs.

Before that date, British pig farmers complained that our higher standards put us at a disadvantage, but pig-rearing is a good example of harmonised conditions. The National Pig Association vows to maintain these welfare standards, saying they 'must not be lowered in pursuit of new trade deals and a post-Brexit cheap food agenda'.[60] It is calling for steps to protect pig producers and consumers, including equivalent standards for meat imports and, if necessary, tariffs and quotas where standards fall short. They want to see strict labelling laws put in place to provide clarity for consumers over differences in production standards. Like many other farmers, they are also calling for free access to the Single Market and EU labour – the cake and eat it version of Brexit.

The trade in pig meat between the UK and other EU countries is largely unrestricted. Like other meat, it's a very different story for pork entering the EU from outside where imports, apart from offal, are subject to sizeable import tariffs. Bone-in hams, shoulders and bellies all attract tariffs between £500-£1,000 per tonne imported, making most imported pork uncompetitive on the EU market, even though the US, Canada and Brazil can produce it more cheaply. The EU and Canada (CETA) agreement[61] proposes that, in seven years' time, free tariffs will be offered for pork from Canada but with a tariff-related quota (TRQ) attached. As we have seen, whether we are bound to CETA, or whether we have to renegotiate anew depends on whether Brexit occurs before or after full ratification of CETA.[62]

Ractopamine is a pig-feed additive used to produce leaner meat. It is banned in the EU, China and Russia, but deemed fit for human consumption in the USA, Canada and Japan. The USDA organic label means no antibiotics or growth hormones, but this doesn't cover Ractopamine as it is neither a growth hormone nor an antibiotic. In 2012, *CODEX Alimentarius* approved the adoption of a maximum residue limit (MRL) of 10 parts per billion (ppb) for muscle cuts of pork, but the EU's EFSA says there is insufficient data to establish an MRL.[63] You can see why the National Pig Association is worried.

Pig products will undoubtedly reach us and drive prices for domestic pork down. When the UK exports to the EU, our pork will be subject to the same EU import tariffs as other third countries, and will become uncompetitive. Given that nearly three-quarters of pork exports are routed through the EU, this could have a dramatic impact.

As part of going global we have an eye on China. Half the world's pigs are in China, so improving pig genetics is high on their agenda. They want to improve both efficiency and impact on the environment. The UK has been exporting frozen and fresh pig semen to China since 2015, in what was claimed to be a £45m deal. This was supposed to benefit the tiny Yorkshire town of Driffield, which specialises in swine semen. However, for various reasons only £50,000 has been shipped so far.[64]

We imported around 140,000 tonnes of pig meat, mostly from the EU, with a quarter of the pork from Denmark. We exported around 40,000 tonnes, again mainly to the EU, although 7,000 tonnes of pork and offal went to China.[65] We will be up against Denmark, as we have not been able to keep pace with Danish modernisation of the curing process and increasing centralisation. Danish pork products carry the wavy 'Danish' brand, which, if sold here, has to be reared in accordance with UK pig standards, particularly regarding space and sows' welfare.[66]

To encourage buying British the UK could decide to impose its own tariffs and SPS standards on imports. Import tariffs could lead to higher pig prices, meaning higher prices for consumers, unless we adopt my suggestion for using subsidies. No tariffs would encourage more imports from outside the EU, threatening the long-term sustainability of UK pig production.

Lamb

The present EU quota system for sheep-meat imports allows New Zealand a lot of tariff-free access. The UK is by far the largest exporter of sheep meat in the EU. France accounts for about half of this. Continued tariff-free access to the EU market will be crucial, but I can't see France welcoming that, as it sees an opportunity to increase its own sheep exports. UK lamb struggles to compete on price outside the EU.[67]

There are 20 tariffs on various bits of sheep under the commodity code 0204 (fresh/frozen sheep meat). Carcasses coming in are taxed at around 40 per cent, with different parts of the carcass taxed at different rates. Lamb is a fine example of the complexities of what may happen if we come out of the Customs Union. The EU tariffs are imposed above set quotas. New Zealand has a higher quota before taxes kick in of 200,000 tonnes; the quota from Australia is 20,000 tonnes. 64,500 tonnes of this New Zealand quota comes to the UK. However, it could be a tricky negotiation over how much of the EU quota is taken by the UK after Brexit. We could alter the quota, so New Zealand lamb is taxed sooner, but that will be contentious. Australia sees the opportunity to export more to the UK.[68]

96 per cent of our lamb exports go to the EU. Without a deal, we would face the 40 per cent Tariff Cliff. When Welsh farmers voted for Brexit – having been promised their subsidies would be safe – nobody spelt out the possible tariff changes should we came out of the Customs Union. Irish sheep farmers will be looking to fill the shortfall of UK lamb going into the EU.

If we want to go global, UK sheep meat struggles to be competitive against the major Southern Hemisphere exporters, who dominate global trade. Exports to these markets are likely to be premium cuts, sold to consumers who are attracted by the UK's pasture-based production systems. Cheaper cuts and offal may find a market opportunity in Asia, the Middle East and emerging economies. We could increase sales (around 3,000 tonnes) of lower-value cuts, as they have virtually no value here. China is already the largest importer of sheep meat and widely expected to increase its imports further, but they already have FTAs with New Zealand and Australia.

We could encourage buying more British, by increasing tariffs or reducing quotas. This could lead to tighter supplies and provide an opportunity for domestic lamb to replace imports, leading to higher prices in the short term, encouraging domestic production to react. UK consumers' preference for legs may limit the extent to which domestic lamb could displace imports.

Our sheep production is highly dependent on a trade deal between the UK and the EU. If UK sheep meat was subject to tariffs when entering the EU, it could result in a collapse in exports. Tariffs could limit movement of Northern Irish lambs to slaughter in Ireland. However, Irish sheep farmers will be licking their lips at the prospect of making up the shortfall of lamb exports to France.

Fish Supper

Brexit was cheered by many in the UK's fishing communities,[69] who blamed the EU for their woes. Over 90 per cent of all fishing communities wanted out.[70] They believed they could catch more fish as a result. Peterhead, the UK's largest white fish and pelagic (mackerel and herring) port, used to license over 120 vessels, but that has fallen to around 20. The slogan 'control our borders' resonated strongly in the UK fishing industry, who queried why land-locked Austria had a bigger fishing quota than the UK. This feeling was aided and abetted by Boris Johnson: 'The EU is pinching our fish.'[71] The UK quota was based on historic catches in the 1970s, when Britain also fished around Iceland, until Iceland extended its own rights to 200 miles in 1976, leading to the 'Cod Wars'.[72]

When doing a deal, the EU will point out that the UK's share of the overall EU fishing catch has grown from the fourth largest catch of any EU country at 652,000 tonnes, to the second largest in the EU by 2014 of 752,000 tonnes.

George Eustice, the Fisheries Minister, says that Britain can instigate a 200-mile Economic Exclusion Zone (EEZ) around Britain. There is precedent for this – Norway has such a zone and it is part of a UN agreement to that effect. Eustice cited UN Article 57,[73] but forgot to mention Article 56 which says: 'In exercising its rights and performing its duties under this Convention in the exclusive economic zone, the coastal State shall have due regard to the rights and duties of other States and shall act in a manner compatible with the provisions of this Convention.' Anything we try to impose will have to be negotiated with the EU if we also want access to their exclusion zone.

The House of Lords said:

> the moment we leave the European Union, the Exclusive Economic Zone (EEZ) will become our exclusive economic zone, exactly as it

> says on the tin. There will be no automatic right for us to fish in other people's EEZs; nor will there be any automatic right for other nation states to fish in ours. We will be excluded immediately, if we have not renegotiated access, from agreements with Iceland, Norway and the Faroes, which are particularly important to our Scottish fleets.[74]

Michael Gove, Secretary of State for the Environment after the 2017 election, went straight to Peterhead to say that there would be a 'sea of opportunity' and that a new approach with a new Fisheries Bill would reduce the number of foreign boats in our waters.[75] Clearly that will be popular with the fishing industry, but we will have to see how well our boats will be received in EU waters.

When considering buying British, there is a curious feature regarding our fish-eating habits. We mainly import the fish we eat – cod and haddock – and export the fish we catch – salmon, scallops, mackerel and crabs.[76]

Until the 1970s, most cod and haddock was caught by British deep-sea fishing fleets in dangerous waters. These fleets were based principally in Grimsby, Hull and Fleetwood and fished the areas round Iceland – until the Cod Wars. Now we import those fish from Iceland, Denmark and Norway. Despite many efforts to persuade people to eat different fish, like herring and mackerel, we still consume getting on for a quarter of a million tonnes of cod. Meanwhile, fleets of boats out of Cornwall export most of their catch of turbot, monkfish, megrim and brill to France and Spain.[77]

When doing any deal, each of the various crustaceans attract a different rate of tax for third countries. Crawfish tails (commodity code 0306111010) imported from outside the EU are subject to a third country duty of 12.50 per cent, whole cooked lobster are subject to 6 per cent while three species of crab and freshwater crayfish are taxed at 7.50 per cent, and several species of prawn at 12 per cent. Once we leave the Customs Union, our own prawn fleets will face this tariff when trying to export into the EU.

Free traders argue that by coming out of the EU, prawns will be cheaper. The 'leave means leave' campaign say Thai prawns would be 42 per cent less, once the 'grip of EU tariffs is prised off non-EU comestibles'.[78] I have a problem with Thai prawns. In December 2004, a massive earthquake under the Indian Ocean led to a tsunami: massive shock waves hitting the coasts of Thailand and Indonesia and washing away an estimated 150,000 people. Obviously there is nothing we can do to stop earthquakes, but mangroves – the trees that grow in sea-water – would have reduced the impact. But Thailand destroyed over 180,000 hectares of mangrove forest to make way for the prawn farms.[79]

We could buy British. The brown shrimp is common in British and Irish waters. They have a shorter, flatter body and are an important part of the food chain. They feed on all sorts of debris, and are eaten by fish and birds, especially in tidal areas. Morecambe Bay potted shrimps are a local delicacy and give the Morecambe football team their name: 'The Shrimps'.

A Blue New Deal launched by the New Economics Foundation seeks to revitalise our coasts and make sure that when restaurants claim to sell local fish, they really are local – not just bought from a local supermarket.[80]

Mineral Water

On Brexit, Directive 2009/54/EC and its related regulation[81] that deals with the marketing and exploitation of natural mineral waters will be up for grabs. Two main types of bottled water are recognised – mineral water and spring water. Mineral water is groundwater that has emerged from the ground and flowed over rock, with treatment restricted to removal of unstable elements such as iron and sulphur compounds, by filtration or decanting. Filtered tap water does not count as 'mineral water' but is 'table water'. A third of all bottled water in the US, complete with labels with mountains and forests, comes from taps. Could the same happen here?

Consumption of bottled mineral water has increased by around 25 times since the 1970s, when Perrier first emerged. This is despite paying about 2,000 times more for it than tap water. This is due in part to scare advertising implying tap water is less safe – despite good UK tap standards.

Figures from the US show that the oil needed to make the plastic bottles required for mineral waters could run a million cars. Every year the average UK household uses 480 plastic bottles, but only recycles 270 of them. Nearly half are not put into recycling facilities, according to Recycle Now, part of the government's waste advisory group WRAP. The rest are incinerated or find their way to landfill or are even shipped abroad.[82] One billion people in the world don't have access to clean water.

In the past, trade unions have campaigned to have drinking fountains at work. Perhaps there should be a new campaign to fill many of those plastic bottles going to waste with our clean drinking water, and send them to people overseas who are desperate for potable drinking water via 'Water Aid'.

Mineral waters and aerated waters that contain added sugar or other sweetenings attract an EU import tariff of 9.6 per cent. We export about $500m worth of flavoured waters, over half of which goes to Ireland. Once we are outside the Single Market, and Ireland remains inside, it will make it harder to export our bottled water, pushing the water up a 10 per cent Tariff Cliff.

Chocolate

As we Brexit, the 'harmonised' EU Directive issued in 2000 about what we can call 'chocolate', may be re-examined.[83] 'It determines the minimum percentage of cocoa butter which can be used. It also determines the possibility to use a quantity of vegetable fats which does not exceed five per cent of the end product.' For years Belgium and France argued that only chocolate made from cocoa butter deserved the name. Post Brexit, it will be possible for UK manufacturers to put as much vegetable fat in chocolate, and still call it chocolate.

When doing any deal, we will have to disentangle ourselves from the current import arrangements. We import the bulk of our chocolate products from the EU, while the EU imports much of the cocoa from West Africa.

According to the FAO: 'Cocoa producing countries limit themselves to mainly exporting beans – rather than manufactured cocoa, or chocolate products – mostly because of tariff escalation. The EU has no tariff for cocoa beans, but a 7.7 per cent and 15 per cent duty on cocoa powder and chocolate crumb containing cocoa butter respectively.'[84]

Going global offers all sorts of prospects. We import $2.4bn worth of chocolate, while we export $888m. Those tariffs allow us to 'add value' to the cocoa beans and sell on as luxury chocolate. Our chocolate industry is worth around $4bn, giving you an idea of the value added to those beans.

The $120bn global chocolate industry relies on ageing trees and small farms in West Africa,[85] which have barely enough money to invest and are under threat from other crops like palm oil. The Ivory Coast is the world's largest producer of cocoa beans – 55 per cent of its cocoa beans are exported to the EU, worth $2bn a year. Only 3 per cent of its cocoa beans are exported directly to the UK, opening up an opportunity to deal directly with the Ivory Coast and other former colonies, like Ghana and Nigeria.

On the other side of the world, where cocoa was first cultivated, there are moves to develop plantations. The company Pacific Agri-Capital are 'developing greenfield cacao plantations in both Colombia and Peru. These countries offer ideal cacao growing conditions with the perfect levels of rainfall, humidity and sunshine. Both countries allow foreign investors to acquire freehold titled land and have superior infrastructure compared to peer cacao growing countries.'[86] The politics of the plantation have not gone away.

Buying British is impossible: we don't grow cocoa. We could take more cocoa directly from West Africa, and reduce the tariffs on cocoa *products,* encouraging African exporters to develop the technologies to add value to their beans and earn better returns for their small farmers.

We can see the actual composition of our chocolate may be different in the future. As you can see in this chapter, every food will be affected differently by Brexit. For some it will be tariffs, others sanitary measures, while many will be involved with import threats or export opportunities. For each food, different discussions and decisions will need to be made. The big questions are, who will be involved and how will the decisions be made in the best interests of the majority of us?

12
What We Can Do

In this chapter, we look at putting the ideas in this book into practice. As we often hear, the proof of the pudding is in the eating, and here I make some predictions as to what may happen – so that we can all check on them in a few years' time. I make proposals as to what we could do – in terms of trade, labour and land, the three areas I prioritised in the introduction. We end with what we can do personally, practically and policy-wise, and set out the principles to guide us in the process.

This is the first labour analysis of food and farming for years. The Brexit vote provoked a debate about food and farming that we have not had for 50 years. Whatever deal is struck, many people will be disappointed. But there will also be many other associated issues, not part of the deal, to discuss and debate. It is not just a matter of laws and trade agreements. Our whole way of life is up for debate. We will want to discuss energy, science, the environment, food, farming, land and labour. Matters that up till now have been the preserve of a few in Brussels are now open to us all.

Our Brexit will stir up long-forgotten matters, but more importantly, it will ask where we want to go. To do that, we will ask ourselves who we are. In this book, we have been looking at this in terms of food and farming, a key part of our culture. We have seen that food production employs more people than any other sector, pays most poorly and creates the most hazardous workplaces. Food and farming use more water than any other sector and while farming puts goodness in, the food-processing sector refines much of that goodness out. Our food needs more and more energy, often drawing on finite fuel sources. Farming uses more land, pollutes more water-courses and reduces biodiversity more than any other sector, while arable farming is destroying the soil. Land-use changes – from forest through pasture to arable – is the second biggest contributor to global warming after the power sector, with nitrate fertilisers and methane adding to that global warming, and in all accounting for 20 per cent of the total GHG contribution. Yet nothing is being done to address this.

We are impacting on the rest of the world, relying on them for half the food we eat. We rely on labour and resources from countries that often can ill afford it. Much of our environmental food footprint occurs overseas. This means we produce two-thirds of our GHGs abroad, and use twice as much land elsewhere to grow our food as we do here. The food we import, often from drought-ridden areas, has consumed and will contain vast amounts of water, while we expect countries overseas to use their water to grow our crops. We use their best land to grow our crops, often under plantation conditions, using spraying techniques we would not allow here.

Simultaneously, we have obesity and food poverty. Inequality in food reflects the inequality in our society. We are producing far more food than is required, resulting in food waste – and obesity. £50bn worth of EU subsidies produce no food. Just as in the rest of the world, where nearly a billion people go hungry, half a million people in the UK now depend on food banks. We do not need more food; we need better food where it is most needed. We have to stop depending on markets to sort matters out – they are the problem, not the solution.

Yet the government stands aside, doing virtually nothing. There is no control over land use. Owners – less than 7 per cent of the population who own 90 per cent of the land – do what they like. The government resists even a soil plan. Farming is excluded from Emissions Trading. There is a proposed sugar tax, but much more is needed to tackle obesity, as current dietary guidelines are clearly not working.

IDENTITY

In the debate about all these food and farm issues, we will raise the issue we didn't talk about in the months leading up to the Brexit referendum – our identity. This compares with the Scottish Referendum, where for over two years the Scots debated what it meant to be Scottish, and as a result are probably clearer about that than we are about being 'British'.

When determining our identity, we should expect differing views. It is not a matter of being British, but also what it means to be English, Welsh Scottish and Irish. We have seen throughout the book that there is no level ploughing field as far as our four nations are concerned.

Scotland is probably clearest in what it may want from a food and farm policy. Food Standards Scotland assumed responsibility in 2015 for advice on nutrition and food labelling. Farming is already a devolved matter, so each nation state should be directly involved in every relevant negotiation, but that won't be straightforward. Already there is debate over who gets which subsidies and what they will do with them. What each country does with the land will vary greatly.

In England, the difference between farming in the East and West will come to the fore. The West may have more in common with Wales and Scotland. We have already seen that only England got rid of its Agricultural Wages Board. But there will be deeper divides.

The response of different regions will also play a part. Each region varies as to how much 'less-favoured land' they have. Previously the EU funded these areas. That will not continue, but how responsive each government can be will depend on how far they are from the hills.

I think food offers us ways to determine our identity like nothing else. Some see our identity in terms of the flag and 'Land of Hope and Glory'. The red, white and blue may have appealed in times gone by, but I think red for labour and green for the environment are a more meaningful combination – land and labour working together. We need to build a new, long-term, sustainable food system, linking our rural lands with our urban growth.

While I hear the PM – any PM – talking about 'shared values', I don't know what is shared. Certainly the land is not. Some look back to a dimly visible golden age, whereas this is the chance to look forward. We can learn about life looking backward, but must live it forward.

Misquoting the Bard of Barking, Brexit 'offers us an opportunity to engage in the debate about what it means to be British on our own terms ... we need to address issues of identity and belonging in the context of equality and rights, using examples from within our own culture.' This is Billy Bragg writing about patriotism.[1] He makes the point that he loves his country and is proud of it in the same way as he loves his son. A good telling off can be good for both. My criticism of what is going on in our food and farm system is based on my love of our land.

We should ensure that we are all better-educated about food, diet and nutrition. We cannot make all our decisions at busy checkouts. Retailers may be beginning to recognise the longer-term consequences of processed foods and ever-lower prices, but are still beholden to consumers who expect cheap food. If ever our government needs to step in, with a programme of inducements, education, research and forward funding, the time is now. But it would seem that there is virtually nothing happening.

We need the state to take control of the land. We need to stop corporates appropriating research we have carried out for a century. We need to grab the initiative from the markets and establish regional control. Markets do not know what is good for us. Adam Smith's invisible hand needs a strong arm behind it. We have a chance to communicate with each other in ways that were unthinkable even a few years ago.

All changes will have impacts. Many workers in the food and farm industries should be better off if the redistribution of subsidies suggested in this book is implemented. For too long, workers in those industries have been threatened by the army of cheap labour. We want to restore respect for the skills needed to look after our land in the long term. Many of us want progress, but few of us like change. So workers in some food industries will feel threatened. At present the whims of markets control what is going to happen. However, we are looking a long way ahead. The changes proposed are not going to happen overnight. For organised workers this is the time to implement 'just transition' arrangements. Just transition has been developed by the trade union movement, and endorsed by the ILO 'to encompass a range of social interventions needed to secure workers' jobs and livelihoods when economies are shifting to sustainable production, including avoiding climate change and protecting biodiversity'.[2] To which we can add health for all – including those most affected by food-related diseases – the poorest.

I remember the general secretary of one of the largest trade unions – friends called him the green giant – who said that when it came to making changes because of environmental impacts, we were not doing our members any favours by pretending they weren't going to happen. I fell out with our local Labour authority when my lecturing union – NATFHE – wanted to make a redundancy agreement. We didn't want redundancies, but we wanted to prepare for them.

Change in food and farming is inevitable. It is how those changes come about that we can influence. We want to influence the changes to the benefit not just of shareholders, but of our members. As workers, by head and hand, we could help drive that. And if in the process, we can have better jobs, better paid, we should go for that. At present, farmworkers are poorly paid because of the demand for cheap food. By redistributing the subsidies I've discussed, we could improve farmworkers' lives *and* improve the health of the nation.

Predictions

Generally

I think it will be a 'dog's Brexit'. We keep being promised 'the best possible deal'. The Prime Minster, when issuing the letter triggering Brexit, said: 'When I sit round the negotiating table in the months ahead. I will represent every person in the whole United Kingdom – young and old, rich and poor, city, town, country, and all the villages and hamlets in between.' I predict she can't – unless she has invented a new substance called 'Miracle May' that enables pigs to fly (tariff free) over the Channel.

Many people will be disappointed with whatever deal is done. Some will be Brexiteers who will be annoyed when they find things haven't changed much – when the fields are still worked by migrant labour. The people in the East of England, who were concerned about migrant fieldworkers, may find different migrant workers, bought in with another version of the scheme like the Seasonal Agricultural Workers Scheme (SAWS), but they will look the same.

In the Northern mill towns, many who also thought immigration would change will be surprised to find it hasn't. I lived in two of them, Blackburn and Nelson, for nearly 40 years. They have yet to find their 'post' mill identity. The fields on their doorstep could offer that – as is starting to happen in Todmorden, an old Yorkshire mill town.

Trade

There will never be a 'level playing field'. The NFU say they want a 'level playing field', meaning a system of trade deals with the rest of the world that takes into account the implicit and explicit subsidies enjoyed by nearly all farmers. But change that phrase to 'level "ploughing" field' and you soon realise there is no such thing. Every bit of land is different. The big landowners will benefit, as they have the best land and the loudest voices.

Throughout history, people throughout the world have been making deals to maximise tax on foodstuffs coming in and out. There are going to be major farm wars, as the free traders try to get rid of tariffs that have been protecting our farms and food producers. Many farmers will feel a very ill wind through the countryside. The war will be between those who want even cheaper food from abroad, and those who want to build our own production.

Trade relations with the EU will take at least five years to sort out, and trade relations with the rest of the world, possibly a generation. What many people don't seem to realise is we can trade any commodity with any country now. We just have to pay tax on it. To get that tax removed, we have to give something in return. For instance, in a trade deal with the US, they would want to ship more corn-based products like HFCS and cornflakes, plus the hormone beef, chlorinated chickens and ractopamine-fed pigs.

Labour

I wish I could say working conditions both on the land and in food services will improve. However, both are threatened. Landworkers could be further threatened by even more imported food, especially in England, without the protection of the AWB. Farm fatality rates will not improve in the next 10 years.

Those in the food-manufacturing and service sectors are likely to be hit by higher prices, because of the slump in the pound, which looks likely to continue. That is why the food-manufacturing sector did not want Brexit. Many restaurants will go out of business because of higher import costs on foodstuffs.

Migrant labour will be harder to come by, and costs of production may increase as a result, but it will still provide the backbone of much farm and food production for the next few years. It will take a generation to build the skills, inducements and investment to rebuild our future on the land.

Land

Land prices may well fall in the short term. Land markets worldwide have struggled against pressure from weak grain prices, which, in cutting farm profits, curb prospects for investment returns. UK land prices have stalled and stuttered since Brexit uncertainty and the possible loss of subsidies, but in the longer term they should be stable.[3]

Sheep and dairy farmers who thought their subsidies were secure won't know what to do. Nobody told them they could be hit from all angles – loss of subsidies, lowered tariff barriers to meat imports, and stiffer tariffs on our meat exports to Europe – presently, the market for most of our meat exports.

I can't see much improvement in UK soils policy, management or monitoring.

The British Aisles

PROPOSALS 1 SOCIETY

Priority – subsidies

Whatever sort of Brexit we have, one thing is certain. The EU subsidies known as CAP will be the responsibility of the UK. They currently make up 50–60 per cent of UK farm income. The government reneged on a statement the Brexit Minister made to Welsh farmers that the subsidies would be maintained. Farmers were so outraged that the Treasurer, Philip Hammond, stepped in to reassure farmers that the subsidies would be safe until 2020. The Tory manifesto extended this to 2022, as we have seen. It is not clear in the Agriculture Bill what levels of support the UK government will be willing to provide beyond this, or whether it will target subsidies in a different way.[4] The time is ripe for a heated debate.

We need to get our hands on £3bn+ CAP funding, to effect changes. We can expect some free traders to argue that we should do away with any subsidies. They will point to New Zealand, which got rid of subsidies in the 1980s. Following the loss of subsidies, there was initially a rise in farm suicides (around 50) and some farmers left to find work elsewhere, but overall there was less migration from the land than expected.[5] New Zealand has much better sheep- and cattle-rearing conditions than the UK. The animals are left to look after themselves, on vast acreages, unlike our sheep-farmers staying up all night lambing. New Zealand has established new markets in China and Japan, whereas we are walking away from our largest and closest market. The parallel with New Zealand doesn't hold up under closer scrutiny.

Nor will there be the EU funds for 'less favourable areas', such as hill farms. Nobody has any idea how these hill farms will be funded. There are already rumblings that there won't be as much subsidy. Subsidies may also decrease in the EU, once our contributions have been removed, although political pressure from rural areas holds greater sway inside the EU. What we do with our own subsidies will be part of the debate over our Identity that we have been discussing.

The NFU proposes 3 cornerstones for a new Domestic Agricultural Policy.[6] They are, with my comments:

1. *Productivity measures and business resilience*. Improving productivity will only make matters worse where there is already overproduction.
2. *Volatility, mitigation measures and management tools*. There will always be volatility in the weather, but reducing monetary volatility would be better dealt with by taking food speculators out of food and farm markets.
3. *Environmental measures*. Until now environmental matters have been dealt with under Pillar 2 of the CAP, which is where agri-environment schemes have been funded. They are part of the EU consultation on CAP at present. However, as we have seen, they do not begin to address the environmental issues around food and farming. There should be separate and distinct funding to improve carbon-saving measures. I have proposed that those CAP subsidies should go to people who work the land. This would not only benefit the 300,000 people who work the land, but could also help to reduce our food trade deficit, stimulate rural economies, help them feed our cities and enhance our own internal food market.

This would do more to bolster our food sustainability and reduce our environmental impacts than anything else. Cutting food imports by half would reduce our overall food footprint by a third, in terms of land use, GHG emissions, water usage and food miles.

We all seem to want cheap food. By subsidising workers, we can keep our home-grown food cheap. And when we buy $33bn worth of our own food, it could carry a 'social care' label, indicating that 10 per cent of each sale goes to our own 'Social Care'. That would bring in the $3bn that Labour calculated we need for Social Care each year. We can do that without any tax increases.

What an incentive for individuals, retailers, manufacturers and food services to buy British. We would be investing in ourselves rather than impoverishing people abroad. We could build rural economies linked with local towns.

We need something that connects the consumer with the worker, rewarding both for buying as locally as possible. We don't have to come out of the EU to change things. If we stay in the EU ...

In and Against

We could stay *in* the EU, without supporting every aspect of EU farming policy. In particular we would still want to change the way 40 per cent of the EU budget goes to the present form of CAP. We could lobby to distribute those subsidies along the lines suggested in the book, redirecting them to landworkers rather than landowners, not only in the UK but throughout Europe.

The EU launched the three-month CAP Consultation in February 2017.[7] While that has now passed, the political process will go on for some time. The EU recognises that a 'growing majority of Europeans consider agriculture and rural areas as important for the future, with more than nine out of ten respondents holding this view.'[8] But I think – or rather, I know – that many people do not agree that CAP should stay in its present form.

Article 39 of the Treaty of Rome specifies that the first two objectives of the Common Agricultural Policy shall be:

(a) To increase agricultural productivity by promoting technical progress and by ensuring the rational development of agricultural production and the optimum utilisation of the factors of production, in particular labour;

(b) Thus to ensure a fair standard of living for the agricultural community, in particular by increasing the individual earnings of people engaged in agriculture.

I believe the suggestions in this book do more than most to meet these objectives. They also respond to changes since the last CAP reform.[9] A report on this consultation has appeared,[10] and if we don't manage to get enough momentum for this round of changes, there will be another round. We want to make sure all our voices are heard, not just those of agribusiness and big farmers, with a permanent lobbying presence in Brussels, throughout what will be a long consultation process. We want younger people to have a voice in the future of our food and farming.

The Conservatives are planning an Agriculture Bill as part of the Brexit process that 'will ensure an effective system is in place to support UK farmers and protect the natural environment after the UK leaves the EU and therefore the Common Agricultural Policy'.[11] The Environment Secretary, Michael Gove, stressed his support for the sector, pledging to 'champion UK farmers' during an NFU reception in the House of Commons. Time will tell how he balances the environmental aspects.

This doesn't sound very different to what the EU is saying. The main conclusions from the consultation are the 'need to strongly support the European model of agriculture and family farming. A reshaped CAP must support farm incomes, deal with market volatility and preserve European agricultural production, also in light of any new trade deals. The income inequality both between rural and urban areas and within the agricultural sector itself should be overcome.'[12]

We need to change the EU CAP, and any UK copies of it, dramatically, along the lines I've suggested in this book. We have the opportunity to do so, whatever happens.

Trade

Whatever form Brexit takes, the big change must be to produce much more of our own food, of many different sorts. We import $66bn worth, and export just half that. Clearly we can't easily produce a lot of food products like bananas and oranges, but we could produce a lot of our vegetables and soft fruit. So, let's aim for importing the same value as we export. That way we could reduce the food trade deficit of $33bn to closer to zero. We should put up tariffs on imported food to encourage home production.

While we doubled food imports in 25 years, we closed down three-quarters of our own land-based research institutions. Let's reverse those trends, and halve imports while we double the size of our research base. While the free marketers are chasing over the world looking for deals for cheap food, we have massive food markets right under our noses – literally.

An old colleague, Joe Hanlon, said ' Just give money to the poor',[13] in the context of development. If you give people money, you create a market from which everything else can develop. Here I mean it somewhat differently. I'm proposing that we give money to our rural workers, that will be spent in our rural economy. We use the £3bn to create a Basic Rural System, using the old subsidies to provide a living wage for all. At present the disincentive is that our food markets are saturated, denying our farmers and food producers their just rewards. If we reduce food imports, the markets are less saturated so farm-gate prices can remain higher, and we use subsidies to keep prices down for customers.

Labour

Chapter 5 set out the main proposal: to subsidise land labour not landowners, thereby putting that money to work. In March 2017, Labour calculated £3bn a year was required for our 'Social Care'. We could pay for those services – without taxing anybody – by buying much more of our own food, creating new markets, through our 'barter code', rewarding healthier food, and hopefully reduce the costs of care on the NHS caused by type 2 diabetes.

We would need a broad alliance to support this proposal. The reds and greens should come together. Labour and environmentalists can unite to make people and the planet healthier. This could be a campaigning issue for farmers, food-manufacturers, the service sector, trade unions, businesses, green NGOs and rural communities. This sort of broad alliance will be needed to challenge the landowners and the free-trade financiers who oppose subsidy changes, which would make the whole operation of food farming more decent, more dignified and better able to deal with the future. This gives a lot of us a distinct voice in the forthcoming debate about what happens to CAP subsidies, and ensuring that food and farm matters are not swamped by other Brexit issues.

There is an idea gaining ground called the 'universal basic income'. Also called basic income guarantee, Citizen's Income, unconditional basic income, or universal basic income, this is a form of social – and food – security promoted by both left and right because welfare recipients are taken away from the paternalistic oversight of conditional welfare-state policies. Similar proposals for 'capital grants provided at the age of majority' date to Thomas Paine's *Agrarian Justice* of 1795.[14] So we create a 'Rural Basic income' and fund from the customer through subsidised and sophisticated barter codes.

Land

We should stop paying anyone to own land and do nothing with it. Thanks to a complex history, subsidies are dispensed according to how much land people own. Check who gets what in farmsubsidy.org.[15]

The Landworkers' Alliance (LWA) presented to DEFRA in London in April 2017 a proposal that future farm support should be separated from landownership, with stricter regulation introduced to ensure farmers get a fairer return from the supply chain.[16] My proposal seems to complement that.

We need to keep environmental schemes separate from farm subsidies. The 'agri-environment' schemes have at best been a mixed success. The aim should be to grow more food here – not necessarily intensively, but sensitively. This will do more to mitigate global warming and look after soils than any other suggestions. In future, conservation schemes should be quite separate and determined locally – in clusters or councils, but not by individual farms. That way we can allocate monies to particular environmental projects which can involve all sorts of partners, rather than relying on individuals.

PROPOSALS 2 SCIENCE

Our land science may struggle to survive following any form of Brexit. Many scientists are bracing themselves for a bumpy landing.[17] UK science has been a magnet for talent and funds, bringing in £7bn between 2007–13, and four times better networked than other EU countries. An Industrial Strategy Challenge Fund of £4.7bn has been set up, but that looks more likely to go to electronics and drugs rather than the land. The UK used to receive £200m/yr from the EU's European Research Council for long-term 'pure' research. We need a coordinated approach, linking universities, colleges, farmers and unions, to make sure we get the funds for research in the future; 'The £160m for Agri Tech for N8 Universities should be the basis, but linked more to local production'. I made a number of suggestions in the relevant chapters on:

Sustainability

We need to grow more of our own foodstuffs to improve our food sustainability, in terms of land use, nitrate fertilisers, soils, GHGs and water and energy use. We should investigate the real extent of carbon loss in arable soils and why it is happening. The Countryside Survey said in 2007 that soil was being mismanaged, yet nothing has been done since then. We need accurate soil indicators and we should develop a soil science to which people can relate – like soil animals – given its importance!

We need more accurate calculations of the possible contribution to global warming of afforestation, and better information about proper versus improper management of moors, pastures and new grazing methods, various arable systems and mixed farming.

Obesity

We should find out why the dietary guidelines are not working, revisit traffic-light labels, recalibrate calorie calculations and introduce a tax on all sugar. Above all, we should aim for a transparent process that examines the evidence and produces recommendations that are then put to the test.

Pesticides

We need to develop clear mandatory Integrated Pest Management guidance, to plan better for future eventualities. Whatever transfer of pesticide control occurs, we should establish a 'Sustainable Use of Pesticides' Committee to monitor use of pesticides *after* approval, and enter into a serious debate over the hazards-based approach versus the risk-assessment approach, in order to ensure that the most appropriate methodology is applied.

But most of all, we should calculate the possible contribution to global warming of herbicide use. I have banged on about each weed being a small carbon-capture and storage unit, but only proper field trials can measure the impacts.

Genetically Modified Organisms

As part of any debate, we should set up a public panel to assess the value of introducing particular GM crops. This should involve various interested parties – businesses, scientists and citizens. Part of that process should include what happens regarding ownership of GM traits discovered through research in public institutions.

PROOF OF THE PUDDING

We can only judge these ideas by seeing them in practice. Table 8 sets out the main possible options with Brexit.

Personal

I've been making the case that the state should step in, because as individuals we may think we have choices, but they are limited. However there is one thing we can do: start a discussion. Whenever you are buying food, whether in a shop or restaurant, ask 'where has this food come from?'

We can all do our bit to ensure that teaching about food and farming becomes part of the curriculum. One in ten 14–16-year-olds in the UK

Photo 10 Most children know little about farming or how food is produced. Credit: Mark Harvey

Table 8 Summary of Possible Changes with Implications for Four Main Aspects of Food and Farming.

Scenarios	*Subsidies*	*Institutions*	*Standards*	*Tariffs*
1. **Stay in** 2. **(and against)**	Change CAP. May be easier to implement plan to subsidise workers not owners. Could go for deal on free movement of labour rather than free movement of people.	Share all existing EU institutions. Save money on replication.	Make less cumbersome, have more say.	N/A
3. **Coming out of EU (e.g. judiciary) but not Single Market or Customs Union**	Probably out of CAP, but keep tariff and non-tariff-free access to EU.	Come out of legal system but stay in financial [system]	Many standards stay same, no power over setting them.	N/A
4. **Outside Single Market**	To control migration, come out of Internal market, so cannot have non-tariff-free access.	Replace many, e.g. pesticides, perhaps share some, e.g. food safety.	Hold on to those we can. Create new ones.	N/A
5. **European Free Trade Area** (e.g. Norway)*	Should maintain similar subsidies but subsidise workers.	Replace all, to ensure healthy access to EU and elsewhere.	More flexible control.	Enter complex tariffs for imports of foods. OK to export to EU – but need special arrangement for foodstuffs.
6. **Out of Single Market, but in CU** (e.g. Turkey)*	Maintain similar subsidies, but subsidise workers not owners.	Create many new ones, use some world institutions.	Watch for lower standards.	Similar tariffs, and OK for 15,000 PAPs.[18] Need to agree agricultural products.
7. **Out of both Single Market and Customs Union, but some 'deal' possible**	Fight for subsidies. Subsidise workers not owners.	Use some world ones, create new ones.	Keep to world, WHO, FAO and WTO standards: *CODEX Alimentarius*	Deal likely to focus on automotive and finance sectors. Mind boggling complex for all agricultural products and even more so for PAPs.
8. **No deal**	Fear that cheap food from overseas as set for sale over high seas.	Use world institutions.	Use world standards like *CODEX Alimentarius.*	WTO say 'start from scratch', which has no precedent.

* Arrangements of these countries with EU exclude agricultural (fish and farming) products.

believes that tomatoes grow underground, while a similar proportion of 8–11-year-olds think pasta comes from an animal. A quarter of primary school children think that cheese comes from plants.[19] Geography could explain how food crops have moved round the world; we could teach science through cooking, using history to recount how the tastes of the rich and poor have changed through the years. Todmorden's Incredible Edible, founded by Pam Warhurst, has made a start on an Edible Curriculum.[20] Every school should have an allotment, with different years taking different responsibilities to provide and cook the food. Schools could collect local seeds and set up school seed banks.[21] Our unions could be working with young people to help develop their growing and cooking skills and to make sure they are properly rewarded later on.

Practical

We need to bring back degrees for horticulture and agriculture, and better reward those with vocational qualifications. The government talks of apprenticeships, but we need much more investment to encourage young people to take up food and farming as a career. We need courses like the Food Entrepreneurs I was involved with at Manchester Metropolitan University, linking cities with the nearby land. In order to do so, access to land is required, so we need to bring back council farms, use brownfield sites, and find growing areas in cities. The states of Minnesota and Michigan, in the US, have both allocated funds of $5m and $30m a year respectively,[22] to develop food production to benefit deserving communities. That is the sort of money UK cities needs to invest. There is an international Food for Cities network, run by the FAO. From Totnes to Todmorden, towns, villages and cities are taking up the challenge. There are nearly 50 such projects that are now part of the Sustainable Cities Network,[23] including Brighton, Bristol and Birmingham.[24] However, their funding is patchy. They require long-term, reliable funding that responds to innovative ideas. That can only come from the state.

Policy

We are just about the only country in the world not to have a food policy. Now is our chance. At the end of the Second World War, there was the

Hot Springs Conference,[25] where all the countries attending agreed to produce as much food as we could. That should be our starting point now. A group of food policy professors has produced a guide 'A Food Brexit: Time to Get Real'[26] that spells out what our government needs to do in terms of policy commitment, a new statutory framework, clear targets, a new Commission, and priorities for negotiating food as part of Brexit.

The retailers have been left in charge for far too long. There are signs they are doing more to encourage buying locally, but they should be doing a lot more – like having shelves full of local grown vegetables for starters. There is discussion about buttons for online shoppers to be able to select a preference for 'British',[27] which could be the start of the barter code. There are many initiatives for interested customers,[28] although I've made it clear more is needed than Union Jack signs. We keep hearing that many who claim to provide local are not. Manufacturers are probably most guilty of buying cheap imports. There needs to be political direction not just voluntary arrangements, whether as inducements or instructions. We simply cannot leave these matters to the supermarkets.

Our future food needs to be diverse, both for biological reasons, but also market ones – so the markets don't get saturated. Here is a chance to change things for the better. We can create new cuisines and new cultures. There are countless TV shows extolling the virtues of local cheeses, ice cream, wines etc. The BBC Food Programme regularly highlights the quality of local produce. Our political parties need to catch up and turn these ideas into practice, creating new markets, and going far beyond farmers' markets.

Lord Larry Whitty, while Labour Under-Secretary of State for DEFRA, organised conferences to encourage public sector procurement of local food. He knew that EU rules of public procurement meant they were supposed to buy the cheapest food through open competition.[29] It is another example of how free-market madness dominates concerns for health and sustainability. He could see that schools, hospitals, prisons and universities, could provide a powerful market to stimulate local food production. These proposals were updated in 2014.[30] When (if) we leave the EU in 2019, the UK government will not be constrained by these laws and could direct public procurement contracts towards healthy local food products, and probably do patients and students a big favour in the process.[31]

Michael Gove, the Environment Minister, could, unfettered by EU law, take up where Lord Larry Whitty left off. As Felicity Lawrence says: 'He could adopt a joined-up policy and target subsidies to increase production of the sort (of food) we need for health – more fruit and vegetables, less sugar and intensive meat production. He could ensure that new trade deals are built on maintaining welfare and environmental standards, rather than lowering them to compete in new markets. He could insist that continued access to foreign labour is tied to the industry, improving what are often appalling working conditions and pay so British workers are drawn back to jobs they now shun.'[32]

Perhaps this cookbook provides recipes for what is needed to make her call come true.

Principles

We need a coalition of red and green, of unions, NGOs, local growing initiatives, businesses and interested political parties, to build a new red-green food economy to challenge power bases.

Here are the basic principles based loosely around what has happened in Cuba, when they were cut off from Russian support.[33]

1. Link rural with urban directly, developing short, direct supply chains to where food is needed. Learn through food and farm networks, developing local production skills and providing vocational training.
2. Bring universities, research Institutes, colleges, extension services and growers together to promote agro-ecology. This is farming based on understanding and working with ecological systems rather than dominating them. We can connect interested groups electronically, so the scientists are not so isolated.
3. Introduce land reform that enables people who want to work the land to have access to land. There are lots of brownfield sites unused, and lots of land going to waste. New 'rights to farm' should be created to requisition land not properly used – as happened during the Second World War – and use it for the good of everybody.
4. Fair prices must be paid to farmers and to farm workers. Markets need to be stabilised by taking in excess production and turning it into other products. More diversification should also be encouraged. The Basic Rural Support, spelt out in this book, could help transform the talk into the walk.

5. Local production must be the order of the day, to reduce energy and transportation, and stop exploiting the land, labour and water of countries overseas. Part of this initiative would be directed to creating local seed banks linked with schools, so that future generations understand their value.

This book has explored the contradiction between free trade and social well-being and how they are ultimately incompatible. There is no such thing as 'free' markets. Everywhere there are taxes and subsidies to hide their deficiencies. We have seen how chasing after free trade goes against food sustainability, food security and social well-being.

We need to stand up for the values of equality and well-being that benefit the many not the few. As both a socialist and a soil zoologist, I believe we should get rid of the magic money trees that benefit the already well off, and plant many more real trees so that everybody can benefit from their fruits. We should be growing all sorts of plants throughout the land, in ways that are not demeaning to workers but promote the pleasures of working the land and save our inheritance. This may sound idealistic, but it is also very practical. We can show what and who we want to be through food. We can do that by cutting out food speculators and investing in our land – once we have 'control over our land'.

List of Abbreviations

AAA Arable Area Aid
ACP Advisory Committee on Pesticides. Replaced now by Expert Committee on Pesticides
ADI Acceptable Daily Intake
ADAS Agricultural Development and Advisory Service
AIAC Agricultural Industry Advisory Committee of HSE (below)
AHDB Agricultural and Horticultural Development Board
AWB Agricultural Wages Board
BAP Biodiversity Action Plan
BBSRC Biotechnical and Biological Sciences Research Council
BSSRS British Society for Social Responsibility in Science
CAP Common Agricultural Policy
CCC Committee on Climate Change
CETA Comprehensive Economic and Trade Agreement (a trade agreement between the EU and Canada)
CLA Country Landowners and Business Association
DEFRA Department of Environment Food and Rural Affairs
DNA Deoxyribose Nucleic Acid
EEA European Economic Area
EEC European Economic Community now EU, the European Union
EFSA European Food Safety Agency
EFTA European Free Trade Association
ETS Emissions Trading Scheme
FAO Food and Agriculture Organisation of the UN
FBDG Food Based Dietary Guidelines
FTA Free Trade Agreement
FTD Food Trade Deficit
FRC Food Research Collaboration
GATT General Agreement on Tariffs and Trade
GHG Greenhouse Gases and GHGE Greenhouse gas emissions
GLA Gangmasters Licensing Authority, now GLAA Gangmasters and Labour Abuse Authority
GEAC Good Agricultural and Environmental Condition

GWP Global Warming Potential
HACCP Hazard analysis and critical control points
HFCS High Fructose Corn Syrup, known in the EU as Isoglucose
HSE Health and Safety Executive
IARC International Agency Research on Cancer
IPCC International Panel on Climate Change
IPM Integrated Pest Management
IUF International Union of Food, Agricultural, Hotel, Restaurant, Catering, Tobacco and Allied Workers' Associations
LWA Landworkers' Alliance
MFN Most Favoured Nation
MRL Maximum Residue Level
NERC Natural Environment Research Council
NFU National Farmers Union
NGO Non-government organisation
NHS National Health Service
NOEL No Observable Effect Level
NUAAW National Union of Agricultural and Allied Workers
NVZ Nitrate Vulnerable Zone
PBR Plant Breeder Rights
OEC Observatory of Economic Complexity
OECD Organisation for Economic Cooperation and Development
PAN Pesticide Action Network
RSPB The Royal Society for Protection of Birds
SAWS Seasonal Agricultural Workers Scheme
SPS Sanitary and Phytosanitary
SUP Sustainable Use of Pesticides
TIFF Total Income From Farming
UN United Nations
USA United States
WHO World Health Organisation
WRAP Waste Resources Action Programme
WTO World Trade Organisation

BOOK OF THE MONTH

Getting Unite members reading

A NOTE TO UNITE TRADE UNION TUTORS, OFFICERS AND REPS

Do you have reps and members asking what books you'd recommend reading?

Unite Education has a large number as since 2013 we have developed a series of reading and history projects.

See BOOK OF THE MONTH at http://www.unitetheunion.org/growing-our-union/education/bookofthemonth/

The subjects chosen have been very varied and include politics, sport, social and labour history and economics. There is a review of each book and details of how to obtain copies.

Unite education has also published some of its own books on famous figures and successful struggles from the union's past including Julia Varley, Tom Jones and Benny Rothman. All are available to download at the link above.

For more details contact
Mark Metcalf on 07392 852561
mcmetcalf@icloud.com
@markmetcalf07

Unite the UNION

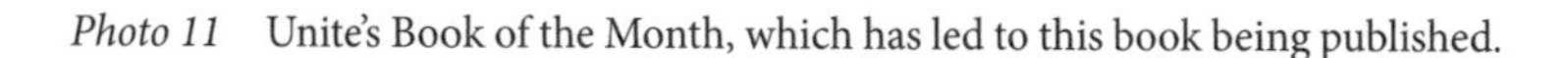
Photo 11 Unite's Book of the Month, which has led to this book being published.

Notes

NB: URL links last accessed August 2017

CHAPTER 1

1. www.nfuonline.com/news/eu-referendum/eu-referendum-news/nfu-council-agrees-resolution-on-the-eu-referendum/
2. www.cla.org.uk/sites/default/files/CLA_Leave_OR_Remain%20Report FINAL.pdf
3. https://tinyurl.com/y9b6xbk6
4. www.bankofengland.co.uk/monetarypolicy/pages/qe/default.aspx
5. http://positivemoney.org/how-money-works/how-banks-create-money/
6. www.ft.com/content/3f9dd17e-47a2-11e7-8d27-59b4dd6296b8
7. www.gov.uk/government/statistics/food-statistics-pocketbook-2015
8. www.diabetes.org.uk/Professionals/Position-statements-reports/Statistics/
9. OEC export/import figures at http://atlas.media.mit.edu/en/
10. www.youtube.com/watch?v=n_wkO4hko70
11. Quoted by Jeremy Seabrook in www.newstatesman.com/2017/04/what-i-learned-about-class-after-my-twin-brother-and-i-were-separated-11-plus
12. www.bbc.co.uk/history/british/victorians/famine_01.shtml
13. www.ecifm.rdg.ac.uk/postwarag.htm
14. www.theguardian.com/global-development/2011/jun/02/us-food-aid-marshall-plan
15. https://fas.org/sgp/crs/misc/R41072.pdf
16. https://open.spotify.com/track/7pWlF5JuhXhya8hMupgi1L Permission to use lyrics from Larrikin Music Pty Limited - /30-432 Carrington Street, - Sydney, NSW, - Australia, 2000).
17. https://sites.google.com/site/whereourfood/
18. P. Palladino (2003), *Plants Patients and the Historian*, New Brunswick, N.J., Rutgers University Press.
19. G. Stapledon (1943), *The Way of the Land*, London, Faber and Faber, p. 171.
20. www.amazon.co.uk/Common-sense-about-starving-world/dp/B0000CLB2F
21. www.fao.org/fileadmin/templates/library/docs/As_I_Recall.pdf
22. https://books.google.co.uk/books/about/The_biology_of_agricultural_systems.html?id=kjVxDoWFzgcC&redir_esc=y
23. www.bssrs.org/groups/agricapital
24. www.bssrs.org/groups/agricapital/bread
25. J. Hanlon, A. Barrientos and D. Hulme (2010) *Just Give Money to the Poor*, Lynne Rienner, Boulder, Co.
26. R. Blatchford (1894) *Merrie England,* London: Clarion Press, pp. 28-9.

27. www.bbc.co.uk/news/business-14473931
28. www.advancingtogether.com/en/home/
29. www.nationalarchives.gov.uk/education/resources/school-dinners/
30. www.express.co.uk/news/uk/81314/EU-s-butter-mountain-costs-taxpayers-236m
31. https://ps.boell.org/sites/default/files/downloads/Perspectives_02-18_Rami_Zurayk4.pdf
32. https://en.wikipedia.org/wiki/Suicide_of_Mohamed_Bouazizi
33. www.theguardian.com/society/2017/may/29/report-reveals-scale-of-food-bank-use-in-the-uk-ifan
34. www.foodaidnetwork.org.uk/

CHAPTER 2

1. www.business-standard.com/article/pti-stories/pub-rules-apply-for-brexit-bill-eu-says-117020701023_1.html
2. www.amazon.co.uk/David-Davis/e/B001HPHO7Y
3. www.gov.uk/government/speeches/environment-secretary-speaks-at-nfu-conference
4. www.gov.uk/government/speeches/environment-secretary-speaks-at-nfu-conference
5. www.parliament.uk/business/committees/committees-a-z/commons-select/environmental-audit-committee/inquiries/parliament-2015/soil-health/
6. www.gov.uk/guidance/eu-protected-food-names-how-to-register-food-or-drink-products
7. www.bbc.co.uk/news/uk-politics-eu-referendum-36575503
8. www.bbc.co.uk/news/uk-wales-politics-36849224
9. www.gpo.gov/fdsys/pkg/BILLS-113hr2642enr/pdf/BILLS-113hr2642enr.pdf
10. https://farmpolicynews.illinois.edu/2017/05/budget-proposal-white-house-seeks-cuts-agriculture/
11. www.reuters.com/article/us-usa-budget-agriculture-idUSKBN18J2TL
12. www.ft.com/content/117f5f10-ed57-11e6-ba01-119a44939bb6
13. http://apps.environment-agency.gov.uk/wiyby/141443.aspx
14. http://c-ipm.org/fileadmin/c-ipm.org/British_NAP__in_EN_.pdf
15. www.hse.gov.uk/contact/faqs/workingtimedirective.htm
16. https://osha.europa.eu/en/legislation/directives/the-osh-framework-directive/the-osh-framework-directive-introduction
17. www.eversheds-sutherland.com/global/en/what/articles/index.page?ArticleID=en/Competition_EU_and_Regulatory/Brexit_and_UK_Environmental_law
18. www.independent.co.uk/news/uk/politics/brexit-eu-regulations-michael-gove-environment-drugs-a7649041.html
19. www.telegraph.co.uk/news/newstopics/eureferendum/12166345/European-referendum-Michael-Gove-explains-why-Britain-should-leave-the-EU.html
20. www.parliament.uk/site-information/glossary/henry-viii-clauses/

21. www.foodsecurity.ac.uk/news-events/news/2017/170214-pr-lessons-from-the-green-peach-aphid.html
22. www.bbc.co.uk/news/business-36083664
23. http://ec.europa.eu/trade/policy/countries-and-regions/countries/turkey/index_en.htm
24. www.bbc.co.uk/news/business-39706224
25. www.farmbusinesssurvey.co.uk/
26. www.ahdb.org.uk/documents/Horizon_Brexit_Analysis_Report-Oct2016.pdf
27. www.wto.org/english/tratop_e/tpr_e/s284_e.pdf
28. www.wto.org/english/res_e/publications_e/wtocan_e.pdf
29. www.wto.org/english/thewto_e/countries_e/european_communities_e.htm
30. www.ft.com/content/5741129a-4510-11e6-b22f-79eb4891c97d
31. www.food.gov.uk/enforcement/regulation/fir/labelling
32. http://madb.europa.eu/madb/sps_product_description_form.htm
33. www.wto.org/english/tratop_e/sps_e/spsund_e.htm
34. www.telegraph.co.uk/news/2017/03/28/theresa-may-fires-starting-gun-brexit-pledge-get-right-deal/

CHAPTER 3

1. www.realclearworld.com/articles/2017/04/17/a_us-uk_trade_agreement_and_the_trump_re-election_campaign_112297.html
2. http://news.bbc.co.uk/1/hi/uk_politics/4287370.stm
3. http://news.bbc.co.uk/1/hi/business/1430770.stm
4. www.gov.uk/government/news/global-ambitions-for-british-food-and-drink-could-boost-economy-by-nearly-3-billion
5. www.gov.uk/government/publications/uk-food-and-drink-international-action-plan-2016-to-2020
6. www.esmmagazine.com/as-markets-bemoan-brexit-brazil-beef-shippers-see-opportunity/29248
7. www.gov.uk/government/speeches/environment-secretary-sets-out-ambition-for-food-and-farming-industry
8. www.marakon.com/insights/strategic-transformation-in-european-food-why-and-how/
9. www.ethicalconsumer.org/ethicalreports/supermarketssectorreport/supermarketsupplychains.aspx
10. www.leeds.ac.uk/news/article/3993/importance_of_british_food_producers
11. www.telegraph.co.uk/comment/9057284/School-holidays-are-a-pointless-relic-of-the-past.html
12. www.manchestereveningnews.co.uk/news/pop-up-fruit-market-heralds-10312812
13. https://sites.google.com/site/seeds4scotland/
14. www2.warwick.ac.uk/fac/sci/lifesci/wcc/gru/events/allender_crop_diversity_fml.pdf
15. www.fao.org/fcit/fcit-home/en/

16. www.appg-agscience.org.uk/linkedfiles/Defra%20food2030strategy.pdf
17. www.morrisons-corporate.com/media-centre/corporate-news/nations-local-foodmakers/
18. www.nfus.org.uk/news/news/lamb-shelfwatch-shows-supermarket-slap-in-the-face-for-scottish-farmers
19. www.fwi.co.uk/business/business-clinic-mob-grazing-tips-success.htm
20. Have a look at my site 'Look at the Land' https://sites.google.com/site/lookattheland/

CHAPTER 4

1. www.gov.uk/government/publications/uk-food-and-drink-international-action-plan-2016-to-2020
2. www.ft.com/content/0a8e05e4-2f2e-11e7-9555-23ef563ecf9a
3. www.tfa.org.uk/tfa-media-release-no-18-farmers-for-action-farming-to-london-march-george-dunns-pre-march-speech/
4. www.theguardian.com/environment/2014/oct/14/ideal-weather-brings-bumper-uk-apple-harvest
5. www.theguardian.com/environment/2014/sep/05/hot-dry-weather-cereal-harvest-british-farmers
6. www.wsj.com/articles/u-s-farm-income-to-fall-to-lowest-level-in-nine-years-1440521337
7. http://economictimes.indiatimes.com/definition/invisible-hand
8. C. Clutterbuck and T. Lang (1982) *More than we can chew* London: Pluto Press.
9. www.flint-global.com/wp-content/uploads/Tacitus-Lecture-2017.pdf
10. www.world-traders.org/2017/04/06/tacitus-debate-2017-2/
11. www.economist.com/blogs/economist-explains/2017/01/economist-explains-4
12. www.wto.org/english/res_e/reser_e/ersd200802_e.pdf
13. https://fsi.stanford.edu/sites/default/files/Agricultural_Trade_Disputes_in_the_WTO.pdf
14. www.firstpost.com/economy/india-us-resolve-food-security-deadlock-heres-what-the-fight-was-about-1981991.html
15. http://uk.businessinsider.com/chancellor-philip-hammond-2017-mansion-house-speech-soft-brexit-austerity-investment-2017-6
16. www.flint-global.com/wp-content/uploads/Tacitus-Lecture-2017.pdf
17. https://tinyurl.com/ycwt90j2
18. www.nature.org/ourinitiatives/regions/southamerica/brazil/explore/brazil-china-soybean-trade.pdf
19. www.economist.com/news/christmas-specials/21636507-chinas-insatiable-appetite-pork-symbol-countrys-rise-it-also
20. www.amazon.co.uk/d/cka/Merchants-Grain-Profits-Companies-Center-Worlds-Supply/0595142109
21. www.oxfam.org/sites/www.oxfam.org/files/rr-cereal-secrets-grain-traders-agriculture-30082012-en.pdf

22. http://thecommonwealth.org/sites/default/files/inline/brexit-opportunities-for-india.pdf
23. https://sites.google.com/site/whereourfood/
24. Profile (1961) Professor Ritchie Calder, *New Scientist*, 9 (216), 30.
25. www.amazon.co.uk/d/Books/When-Rivers-Journeys-Heart-Worlds-Water-Crisis/1552637417/ref=pd_sbs_14_img_1?_encoding=UTF8&psc=1&refRID=PG7W5WJ1R02VK4T8V6T3
26. www.futures-investor.co.uk/futures_trading_introduction.htm
27. www.fao.org/news/story/en/item/150900/icode/
28. www.globaljustice.org.uk/sites/default/files/files/resources/food_spec_statement_02.2011_0.pdf
29. www.theglobeandmail.com/report-on-business/food-speculation-spurs-us-trading-crackdown/article674061/
30. http://uk.reuters.com/article/2011/06/15/uk-britain-farming-minister-idUKTRE75E2AS20110615
31. www.globaljustice.org.uk/food-speculation
32. www.nfuonline.com/sectors/crops/crops-news/nfu-influences-eu-directive-on-futures-market/
33. https://research.rabobank.com/far/en/sectors/regional-food-agri/weighing-up-future-food-security-in-the-uk.html
34. www.channel4.com/info/press/news/supermarkets-brexit-and-your-shrinking-shop-channel-4-dispatches
35. https://blogs.spectator.co.uk/2017/02/brexit-mean-cheaper-food-dont-open-prosecco-yet/#
36. https://research.rabobank.com/far/en/sectors/regional-food-agri/weighing-up-future-food-security-in-the-uk.html
37. www.gov.uk/guidance/e-exporting
38. www.goodreads.com/book/show/1248039.The_Seven_Cultures_Of_Capitalism

CHAPTER 5

1. http://www.iuf.org/w/
2. www.bbc.co.uk/news/business-40354331
3. www.independent.co.uk/news/uk/politics/hard-brexit-leaving-eu-britain-consequences-effects-dangers-food-supply-uk-supermarkets-warning-a7448536.html
4. www.seasonalberries.co.uk/2017/06/latest-news/summer-fruit-prices-set-to-soar-as-a-result-of-brexit-urgent-action-needed-to-ensure-essential-workforce.html
5. https://yougov.co.uk/news/2017/03/29/attitudes-brexit-everything-we-know-so-far/
6. http://theconversation.com/mcdonalds-concession-on-zero-hours-contracts-is-a-boost-for-unions-and-the-labour-party-76734

7. www.moypark.com/en/news/moy-park-launches-grower-expansion-programme
8. www.unitetheunion.org/uploaded/documents/landworker01020911-8412.pdf
9. www.youtube.com/watch?v=yavMtq1bLwI
10. www.theguardian.com/uk/2007/sep/24/labourconference.workandcareers
11. www.ethicaltrade.org/eti-base-code
12. www.ethicalgrowers.org.uk/
13. www.theguardian.com/uk/2007/sep/24/labourconference.workandcareers
14. www.publications.parliament.uk/pa/cm201617/cmselect/cmenvfru/1009/100904.htm#footnote-026
15. https://tinyurl.com/yabrfevj
16. https://tinyurl.com/y768nemf
17. P.P. Courteney (1965) *Plantation Agriculture*, London: Bell & Sons.
18. E. Hyams (1976) *Soil and Civilisation*, London: Thames & Hudson, p. 234.
19. https://en.wikipedia.org/wiki/Nikolai_Vavilov
20. http://semantics.uchicago.edu/kennedy/classes/sum07/myths/creoles.pdf
21. http://humanityunited.org/pdfs/Modern_Slavery_in_the_Palm_Oil_Industry.pdf
22. www.gov.uk/government/publications/modern-slavery-strategy
23. http://services.parliament.uk/bills/2014-15/modernslavery.html
24. www.foodmanufacture.co.uk/People/Eliminate-slaves-from-food-and-drink
25. www.jrf.org.uk/work/forced-labour
26. www.freshproduce.org.uk/newsdesk/fpc-updates/2017/june/glaa-prosecution/
27. www.foodethicscouncil.org/uploads/publications/businessforum181108.pdf
28. See Home Office video on Modern Slavery www.youtube.com/watch?v=Jv1H_fAoOG4
29. P.P. Courteney (1965), *Plantation Agriculture*, London: Bell & Sons, p. 52 and p. 126.
30. www.ilo.org/dyn/normlex/en/f?p=NORMLEXPUB:12100:0::NO::P12100_ILO_CODE:P110
31. http://foodresearch.org.uk/wp-content/uploads/2016/07/Agricultural-labour-briefing-FINAL-4-July-2016.pdf
32. www.publications.parliament.uk/pa/cm201011/cmhansrd/cm111025/debtext/111025-0002.htm
33. www.youtube.com/watch?v=5bevi4Hsr8I
34. www.publications.parliament.uk/pa/ld201213/ldhansrd/text/130306-0002.htm#13030690000079
35. www.publications.parliament.uk/pa/ld201213/ldhansrd/text/130306-0002.htm#13030690000079 Col 1557
36. https://tinyurl.com/y7fypsjw
37. www.walesonline.co.uk/news/wales-news/agriculture-wages-bill-uk-government-7394743
38. www.gov.scot/Resource/0049/00491331.pdf
39. www.hse.gov.uk/foi/fatalities/2009-10.htm

40. https://en.wikipedia.org/wiki/British_Forces_casualties_in_Afghanistan_since_2001
41. See my website, www.epaw.co.uk/farm.html as to how we did it and my free online learning resources at level 2 www.epaw.co.uk/level2 and level 3 www.epaw.co.uk/level3/introd.html.
42. www.sciencedirect.com/science/article/pii/S0048969707011564
43. https://sites.google.com/site/pesticideexposure/pan-unite-survey
44. www.publications.parliament.uk/pa/cm201617/cmselect/cmenvfru/1009/100902.htm
45. www.publications.parliament.uk/pa/cm201617/cmselect/cmenvfru/1009/100906.htm#_idTextAnchor011
46. www.ahdb.org.uk/documents/Horizon_Brexit_Analysis_20Sept2016.pdf
47. www.ahdb.org.uk/news/AHDbEUReferendum200916.aspx and www.ahdb.org.uk/documents/Horizon_Brexit_Analysis_20Sept2016.pdf
48. BBC *Farming Today* (16 May 2017).
49. www.resolutionfoundation.org/media/press-releases/brexit-vote-requires-migrant-reliant-firms-to-rethink-business-models/
50. www.silsoeresearch.org.uk/sri-history/sri-history.html
51. www.gov.uk/government/speeches/agricultural-tractor-and-trailer-weight-and-speed-limit-regulations
52. www.ft.com/content/3f9dd17e-47a2-11e7-8d27-59b4dd6296b8
53. www.ahdb.org.uk/documents/Horizon_Brexit_Analysis_20September2016.pdf
54. www.food.gov.uk/business-industry/food-hygiene/haccp
55. www.haccpforexcellence.com/home/history_of_haccp
56. http://fabflour.co.uk/fab-bread/the-chorleywood-bread-process-2/
57. www.unitetheunion.org/uploaded/documents/0216-FDA%20Strategy%20doc11-27547.pdf
58. www.notable-quotes.com/m/marx_karl.html#ZXXOzWbxDcGoFwql.99

CHAPTER 6

1. www.nfuonline.com/misc/press-centre/press-releases/withdrawal-of-soil-framework-directive-welcomed/
2. http://ec.europa.eu/environment/soil/process_en.htm
3. www.newstatesman.com/life-and-society/2011/03/million-acres-land-ownership
4. https://sites.google.com/site/sustainablefoodreports/2014/garden-controversy
5. K. Cahill (2006) *Who owns the world*, London: Mainstream Publishing.
6. www.newstatesman.com/politics/2015/06/leader-case-taxing-land
7. Alfred Russel Wallace (1882) *Land Nationalisation*, London: W. Reeves.
8. www.britannica.com/biography/George-Stapledon
9. https://ec.europa.eu/agriculture/envir/measures_en
10. *Assessment of the effects of Environmental Stewardship on landscape character* (2014) Natural England Commissioned Report 158.

11. www.see.leeds.ac.uk/fileadmin/Documents/research/sri/briefingnotes/SRIBNs-4.pdf
12. Letters (2016) *Observer* 11 September.
13. https://raptorpersecutionscotland.wordpress.com/2016/04/29/rspb-complaint-sparks-european-legal-action-over-grouse-moor-burning/
14. http://europa.eu/rapid/press-release_MEMO-17-1045_en.htm
15. www.rspb.org.uk/our-work/our-positions-and-campaigns/campaigning-for-nature/casework/details.aspx?id=tcm:9-326701For latest, see local blog www.energyroyd.org.uk/archives/category/floods
16. www.woodlandtrust.org.uk/publications/2013/02/the-pontbren-project/
17. www.forestry.gov.uk/readreport
18. www.forestry.gov.uk/pdf/InternetCCAP.pdf/$file/InternetCCAP.pdf
19. www.forestry.gov.uk/forestry/infd-8j3jv7
20. www.forestry.gov.uk/forestry/infd-8macqb
21. www.forestry.gov.uk/pdf/fcfc124.pdf/$file/fcfc124.pdf
22. www.forestry.gov.uk/pdf/6_planting_more_trees.pdf/$FILE/6_planting_more_trees.pdf
23. www.confor.org.uk/media/246261/thriving-forestry-and-timber-sector-in-a-postbrexit-world-oct2016.pdf
24. http://webarchive.nationalarchives.gov.uk/+/www.hm-treasury.gov.uk/sternreview_index.htm Annex 7
25. http://onlinelibrary.wiley.com/doi/10.1111/gcbb.12357/pdf
26. https://ec.europa.eu/energy/en/topics/renewable-energy/biomass
27. www.treco.co.uk/resources/how-to-guides/save-money-on-fuel-with-miscanthus
28. http://adlib.everysite.co.uk/resources/000/023/838/miscanthus-guide.pdf
29. www.biofuelwatch.org.uk/2013/biomass-faq-2/#C3
30. www.biofuelwatch.org.uk/2013/biomass-faq-2/#C11
31. www.biofuelwatch.org.uk/2013/biomass-faq-2/#C11
32. www.fwi.co.uk/arable/three-crop-rule-sparks-fears-over-arable-profits.htm
33. www.euractiv.com/section/uk-europe/news/uk-promises-scrapping-three-crops-farm-rule-post-brexit/
34. www.gov.uk/government/uploads/system/uploads/attachment_data/file/69261/pb13297-soil-strategy-090910.pdf
35. www.independent.co.uk/news/uk/home-news/soilerosion-a-major-threat-to-britains-food-supply-says-government-advisory-group-10353870.html
36. www.theccc.org.uk/2015/06/30/urgent-action-needed-now-to-avoid-increasing-costs-and-impacts-of-climate-change-in-the-uk/
37. www.theccc.org.uk/wp-content/uploads/2015/06/Cranfield-University-for-the-ASC.pdf
38. www.gov.uk/government/uploads/system/uploads/attachment_data/file/228842/8082.pdf
39. www.gov.uk/government/publications/safeguarding-our-soils-a-strategy-for-england
40. www.countrysidesurvey.org.uk/sites/www.countrysidesurvey.org.uk/files/CS_UK_2007_TR9-revised%20-%20Soils%20Report.pdf

41. https://en.wikipedia.org/wiki/Agriculture_in_the_United_Kingdom
42. www.answers.com/Q/How_much_does_one_acre_inch_of_top_soil_weigh
43. www.forestry.gov.uk/pdf/6_planting_more_trees.pdf/$FILE/6_planting_more_trees.pdf
44. www.publications.parliament.uk/pa/cm201617/cmselect/cmenvaud/180/180.pdf (p 19)
45. www.theccc.org.uk/wp-content/uploads/2015/06/6.737_CCC-BOOK_WEB_250615_RFS.pdf Chapter 5
46. www.theccc.org.uk/wp-content/uploads/2015/06/Annex-5-Agriculture.pdf
47. https://link.springer.com/article/10.1023/A%3A1009766510274
48. www.fwi.co.uk/arable/check-the-rules-on-ploughing-up-grassland.htm
49. www.parliament.uk/business/committees/committees-a-z/commons-select/environmental-audit-committee/news-parliament-2015/soil-health-report-published-16-17/
50. www.farming.co.uk/news/article/12939
51. www.nrcs.usda.gov/wps/portal/nrcs/main/soils/health/assessment/
52. www.sheffield.ac.uk/news/nr/allotments-could-be-key-sustainable-farming-1.370522
53. http://randd.DEFRA.gov.uk/Default.aspx?Menu=Menu&Module=More&Location=None&Completed=1&ProjectID=17595
54. www.nrcs.usda.gov/wps/portal/nrcs/detail/soils/edu/college/?cid=nrcs142p2_054312
55. www.foodethicscouncil.org/blog/82/19/Save-our-Soils/
56. www.soilhealth.see.uwa.edu.au/components/animals
57. http://jncc.DEFRA.gov.uk/page-5155
58. www.quizrevolution.com/QR2/ch/a238801/
59. www.soilanimals.com/look/uk-soil-survey
60. www.sciencedirect.com/science/article/pii/S0262407909622077
61. www.youtube.com/watch?v=ONQFdGZGwIw
62. www.youtube.com/watch?v=Tsa2Mtn5CsU
63. www.janvanduinen.nl/collembolaengels.html
64. www.youtube.com/watch?v=AMBgEioyaV8
65. https://drive.google.com/file/d/0B_nKNzHmp2VjbmxOT1cyaDVXeVE/view
66. https://sites.google.com/site/soilanimals/learn/global-warming-and-soil-animals
67. www.soilanimals.com/learn/global-warming-and-soil-animals
68. www.soilanimals.com/dig-deeper/birth-of-the-earth
69. http://ec.europa.eu/environment/soil/process_en.htm
70. https://sites.google.com/site/soilanimals/get-involved/petition
71. http://4p1000.org/understand
72. https://sites.google.com/site/soilanimals/news/international-year-of-the-soils
73. www.parliament.uk/business/committees/committees-a-z/commons-select/environmental-audit-committee/inquiries/parliament-2015/soil-health/
74. www.parliament.uk/business/committees/committees-a-z/commons-select/environmental-audit-committee/inquiries/parliament-2015/soil-health/

75. www.gov.uk/guidance/standards-of-good-agricultural-and-environmental-condition
76. www.soilassociation.org/media/4671/runaway-maize-june-2015.pdf
77. http://assurance.redtractor.org.uk/contentfiles/Farmers-5496.pdf
78. http://publications.naturalengland.org.uk/publication/6432069183864832
79. www.landis.org.uk/data/nsi.cfm
80. www.gutenberg.org/ebooks/3754 p.291
81. www.gov.uk/government/organisations/natural-england/about/research#overview
82. www.landis.org.uk/overview/referencecentre.cfm
83. www.plymouthherald.co.uk/government-initiative-r-amp-d-innovation-hole/story-19592702-detail/story.html
84. http://n8agrifood.ac.uk/
85. www.sustainablefood.com/efra.htm
86. www.publications.parliament.uk/pa/cm200809/cmselect/cmenvfru/213/213i.pdf
87. https://sites.google.com/site/ukfoodsupply/land-science
88. https://niabarchive.wordpress.com/?s=sold&submit=
89. https://sites.google.com/site/ukfoodsupply/land-science/wellesbourne
90. www.sustainablefood.com/wellesbourne.html
91. www.prospect.org.uk/news/id/2009/November/18/Lab-closure-is-%E2%80%98scientific-vandalism%E2%80%99-says-union
92. www2.warwick.ac.uk/fac/sci/lifesci/wcc/cropcentre/
93. www2.warwick.ac.uk/fac/sci/lifesci/wcc/gru/
94. www.kentonline.co.uk/kent-business/county-news/east-malling-research-rescue-niab-90657/
95. www.ahdb.org.uk/news/documents/AHDBResponseonBISStrategyforAgrTech20Nov2012.pdf p11
96. https://royalsociety.org/topics-policy/publications/2009/reaping-benefits/
97. www.ukbap.org.uk/how-brexit-will-affect-biology-employment/

CHAPTER 7

1. https://ec.europa.eu/europeaid/sectors/food-and-agriculture/sustainable-agriculture-and-rural-development/agricultural-development_en
2. www.coventry.ac.uk/research/areas-of-research/agroecology-water-resilience/
3. www.futureoffood.ox.ac.uk/sustainable-intensification
4. http://rsif.royalsocietypublishing.org/content/13/114/20151001
5. http://webarchive.nationalarchives.gov.uk/+/www.hm-treasury.gov.uk/sternreview_index.htm Annex 7
6. http://webarchive.nationalarchives.gov.uk/+/www.hm-treasury.gov.uk/sternreview_index.htm Annex 7
7. https://blogs.scientificamerican.com/plugged-in/10-calories-in-1-calorie-out-the-energy-we-spend-on-food/
8. https://tinyurl.com/y8gsouqr

9. www.sustainablefood.com/outlooks.pdf
10. http://publications.jrc.ec.europa.eu/repository/bitstream/JRC96121/ldna27247enn.pdf
11. www.fcrn.org.uk/
12. http://grantham.sheffield.ac.uk/how-to-reduce-the-environmental-impact-of-a-loaf-of-bread/
13. www.sustainablefood.com/guide/foodemissionspolicy.html
14. http://ec.europa.eu/agriculture/envir/index_en.htm
15. www.irishtimes.com/news/environment/coveney-claims-major-breakthrough-on-eu-agriculture-emissions-talks-1.1639934#.UsEQlSrEQkY.twitter
16. www.amazon.com/When-Rivers-Run-Dry-Water/dp/0807085723
17. www.foodethicscouncil.org/blog/5/19/Water-risks-a-stewardship-approach/
18. http://en.wikipedia.org/wiki/Virtual_water
19. www.gdrc.org/uem/footprints/water-footprint.html
20. http://waterfootprint.org/media/downloads/Kenya_Water_Footprint_Profile1_1.pdf
21. https://en.wikipedia.org/wiki/Water_conflict
22. http://staging.unep.org/yearbook/2010/
23. www.sustainablefood.com/guide/nitrogenissue.html
24. www.publications.parliament.uk/pa/cm200506/cmhansrd/vo060424/text/60424w01.htm
25. www.sustainablefood.com/guide/nitrogenissue.html
26. www.publications.parliament.uk/pa/cm200506/cmhansrd/vo060310/debtext/60310-18.htm
27. www.gov.uk/guidance/nutrient-management-nitrate-vulnerable-zones
28. www.npa-uk.org.uk/New_NVZ_designations_for_England_announced.html
29. www.food.gov.uk
30. www.fao.org/docrep/007/y5609e/y5609e01.htm
31. www.bbc.co.uk/news/uk-england-kent-19573885
32. www.countryfile.com/countryside/truth-about-pears
33. www.commonground.org.uk/an-apples-orchards-gazetteer/
34. http://jncc.DEFRA.gov.uk/page-5717
35. www.lovefoodhatewaste.com/why-save-food
36. www.wrap.org.uk/content/quantification-food-surplus-waste-and-related-materials-supply-chain
37. www.wrap.org.uk/category/initiatives/courtauld-commitment
38. http://foodandpoverty.org.uk/
39. http://feedbackglobal.org/campaigns/gleaning-network/
40. www.crowdfunder.co.uk/the-larder
41. www.thegrocer.co.uk/home/topics/waste-not-want-not/how-france-is-leading-the-way-on-food-waste/536447.article
42. https://sustainabledevelopment.un.org/topics
43. https://ec.europa.eu/food/safety/food_waste/eu_actions_en
44. https://agenda.weforum.org/2015/07/countries-emitting-most-greenhouse-gas/

45. http://feedbackglobal.org/food-waste-scandal/
46. http://journals.plos.org/plosone/article?id=10.1371/journal.pone.0165797
47. www.elementascience.org/articles/10.12952/journal.elementa.000116/

CHAPTER 8

1. http://ec.europa.eu/eurostat/statistics-explained/index.php/Overweight_and_obesity_-_BMI_statistics
2. http://ec.europa.eu/health/nutrition_physical_activity/policy/strategy_en
3. http://obesity.thehealthwell.info/search-results/eu-action-plan-childhood-obesity-2014-2020?source=relatedblock
4. www.thegrocer.co.uk/finance/brexit/childhood-obesity-strategy-under-threat-from-brexit-result/538162.article
5. https://tinyurl.com/ya6owepg
6. www.thedrum.com/news/2016/08/19/sainsbury-s-boss-mike-coupe-slams-theresa-may-s-childhood-obesity-plan
7. http://thelancet.com/journals/lancet/article/PIIS0140-6736(11)60814-3/abstract
8. www.diabetes.org.uk/Professionals/Position-statements-reports/Type-2-diabetes-prevention-early-identification/
9. www.sciencedaily.com/releases/2014/03/140317174502.htm
10. http://ajcn.nutrition.org/content/early/2010/01/13/ajcn.2009.27725.short
11. www.nationalobesityforum.org.uk/index.php/_news_/746-nt9D.html
12. www.theguardian.com/society/2016/may/28/national-obesity-forum-advice-fat-dangerous, www.sciencemediacentre.org/expert-reaction-to-new-report-on-diet-as-published-by-the-national-obesity-forum/
13. www.newscientist.com/article/mg23030771-600-carb-your-enthusiasm-are-bread-pasta-and-spuds-making-you-fat/
14. http://en.wikipedia.org/wiki/Seven_Countries_Study
15. www.sevencountriesstudy.com/about-the-study/
16. http://aje.oxfordjournals.org/content/early/2012/11/19/aje.kws374.full
17. https://authoritynutrition.com/modern-nutrition-policy-lies-bad-science/
18. www.epi.umn.edu/cvdepi/essay/famous-polemics-on-diet-heart-theory/
19. https://en.wikipedia.org/wiki/Seven_Countries_Study#cite_note-43
20. T. Cleave (1969) *Diabetes, Coronary Thrombosis and the Saccharine Disease*, Wright: Bristol.
21. www.myauz.com/1here/Goodper cent20Calories,per cent20Badper cent20Caloriesper cent20-per cent20Garyper cent20Taubes.pdf
22. J. Yudkin (1972) *Pure, White and Deadly*, Viking: New York, p. 85.
23. www.theguardian.com/society/2016/apr/07/the-sugar-conspiracy-robert-lustig-john-yudkin
24. www.penguin.co.uk/books/9417/pure-white-and-deadly/
25. https://authoritynutrition.com/6-graphs-the-war-on-fat-was-a-mistake/
26. https://en.wikipedia.org/wiki/Earl_Butz, www.kingcorn.net/
27. https://tinyurl.com/ybhc8kyq

28. www.diabetes.co.uk/in-depth/every-last-shred-evidence-low-fat-dietary-guidelines-never-introduced/
29. http://openheart.bmj.com/content/2/1/e000196
30. http://webarchive.nationalarchives.gov.uk/20170110171021/www.noo.org.uk/NOO_about_obesity/adult_obesity/UK_prevalence_and_trends
31. www.nhs.uk/Livewell/Goodfood/Pages/Eat-less-saturated-fat.aspx
32. www1.msjc.edu/hs/nutr100/energy_macro_need.html
33. https://thedietitianspantry.com/2015/04/01/dietary-guidelines-around-the-world/
34. https://justmeint1health.wordpress.com/tag/wistar-institute/
35. www.gatsby.ucl.ac.uk/~pel/fat/fat1.html
36. http://health.gov/dietaryguidelines/2015/guidelines/
37. https://tinyurl.com/nsy4tkw
38. https://sites.google.com/site/sustainablefoodreports/2016/new-us-dietary-guidelines
39. www.eufic.org/article/en/expid/food-based-dietary-guidelines-in-europe/
40. https://tinyurl.com/y7emxbav
41. www.newstatesman.com/politics/health/2016/06/travel-back-time-1970s-britain-and-see-yourself-why-people-were-so-slim
42. https://authoritynutrition.com/6-reasons-why-a-calorie-is-not-a-calorie/
43. www.scientificamerican.com/article/how-do-food-manufacturers/
44. For details see my web page 'A fat coke please' https://sites.google.com/site/sustainablefoodreports/fat-coke
45. www.theguardian.com/books/2013/feb/24/salt-sugar-fat-moss-review
46. http://ec.europa.eu/agriculture/sites/agriculture/files/sugar/doc/sugar-faq_en.pdf

CHAPTER 9

1. www.fginsight.com/news/news/what-will-pesticides-regulation-look-like-post-brexit-15907
2. www.ahdb.org.uk/documents/Horizon_Brexit_Analysis_january2017.pdf
3. http://en.wikipedia.org/wiki/Neonicotinoid#cite_note-29
4. www.nature.com/news/the-buzz-about-pesticides-1.11626?dm_i=1ANQ,10QP7,6LPVZ3,3444M,1
5. researchbriefings.files.parliament.uk/documents/SN06656/SN06656.pdf
6. https://sites.google.com/site/sustainablefoodreports/neonics/background
7. https://sites.google.com/site/sustainablefoodreports/neonics/acp
8. www.thetimes.co.uk/article/brussels-and-its-busy-bees-are-a-perfect-pest-gx2q3vdxcx7
9. www.farming.co.uk/news/article/10448
10. http://rspb.royalsocietypublishing.org/content/282/1819/20152110
11. www.newscientist.com/article/2098858-controversial-pesticides-may-be-lowering-the-sperm-count-of-bees/
12. www.nature.com/articles/ncomms12459

13. www.gov.uk/government/uploads/system/uploads/attachment_data/file/610174/ecp-ministers-advice-1704.pdf
14. www.rothamsted.ac.uk/news/rothamsted-questions-eu-pesticide-ban-chemicals-industry-eyes-brexit-breakthrough-bees
15. https://en.wikipedia.org/wiki/Integrated_pest_management
16. http://ec.europa.eu/food/plant/pesticides/sustainable_use_pesticides/index_en.htm
17. www.gov.uk/government/publications/pesticides-uk-national-action-plan
18. https://tinyurl.com/ycat5dh9
19. www.voluntaryinitiative.org.uk/en/vi-schemes/ipm-plans
20. www.voluntaryinitiative.org.uk/en/home
21. www.hse.gov.uk/aboutus/meetings/iacs/aiac/020507/minutes.pdf
22. www.adas.uk/News/brexit-and-implications-for-pesticide-approvals-and-use
23. www.nfuonline.com/news/featured-article/back-british-farming-brexit-and-beyond-the-nfu-20/#top
24. http://jech.bmj.com/content/early/2016/03/03/jech-2015-207005
25. https://en.wikipedia.org/wiki/List_of_IARC_Group_2A_carcinogens
26. https://corporateeurope.org/food-and-agriculture/2015/11/efsa-and-member-states-vs-iarc-glyphosate-has-science-won
27. https://tinyurl.com/yamjqe7t
28. http://jech.bmj.com/content/early/2017/02/22/jech-2016-208463
29. https://sites.google.com/site/soilanimals/news/gardeners-world
30. www.soilanimals.com/get-involved/diy
31. https://sites.google.com/site/savewyecollege/
32. J. Cook and C. Kaufman, *Portrait of a Poison* London: Pluto Press.
33. www.politico.eu/article/Glyphosate-weedkiller-decision-adds-to-brexit-momentum-for-uk-farmers/
34. https://ec.europa.eu/food/plant/pesticides/max_residue_levels/eu_rules_en
35. www.fao.org/fao-who-codexalimentarius/standards/pestres/en/
36. https://tinyurl.com/ydd3sluj
37. https://tinyurl.com/y7e56u6r
38. www.ahdb.org.uk/documents/Horizon_Brexit_Analysis_january2017.pdf

CHAPTER 10

1. http://news.bbc.co.uk/1/hi/special_report/1999/02/99/food_under_the_microscope/285408.stm& www.princeofwales.gov.uk/media/speeches/article-hrh-the-prince-of-wales-titled-questions-about-genetically-modified-organisms
2. http://news.bbc.co.uk/1/hi/uk/298229.stm
3. www.theecologist.org/News/news_analysis/1745491/the_gm_lobby_and_its_seven_sins_against_science.html
4. www.cell.com/current-biology/abstract/S0960-9822(13)00513-7
5. www.bbc.co.uk/news/science-environment-25885756
6. www.norfolkplantsciences.com/
7. www.bigbarn.co.uk/blog/2014/01/27/are-gm-tomatoes-healthy/

8. www.pendlewitches.co/
9. www.bbc.co.uk/news/science-environment-25885756
10. http://en.wikipedia.org/wiki/Kumato#cite_note-1
11. http://en.wikipedia.org/wiki/Kumato#cite_note-3, www.facebook.com/kumatotomatoes
12. www.ers.usda.gov/webdocs/publications/42517/13616_aib786_1_.pdf?v=41055
13. www.nytimes.com/2010/02/14/magazine/14Biology-t.html?pagewanted=all&_r=0
14. www.linkedin.com/groups/1691777/1691777-6103626083259731969
15. www.syngenta-growth.com/en/home/
16. www.syngenta.com/global/corporate/en/goodgrowthplan/home/Pages/homepage.aspx
17. www.statista.com/statistics/257489/revenue-of-top-agrochemical-companies-worldwide-2011/
18. www.iied.org/sustainable-agriculture-china-then-now
19. https://sites.google.com/site/sustainablefoodreports/2016/china-crisis
20. www.birthdefects.org/agent-orange/
21. www.euronews.com/2017/03/30/bayer-monsanto-merger-is-marriage-made-in-hell-say-activists
22. www.rothamsted.ac.uk/our-science/rothamsted-gm-wheat-trial
23. www.rothamsted.ac.uk/news/gm-plants-promise-fish-oils-aplenty
24. www.bbc.co.uk/news/science-environment-26189722
25. https://geofftansey.wordpress.com/2012/08/07/non-gm-blight-resistant-potatoes-champion-the-sarvari-research-trust-faces-collapse-2/
26. www.fooddive.com/news/usda-oks-genetically-modified-potato/330750/
27. www.food.gov.uk/science/acrylamide-0
28. www.sustainablefood.com/guide/acrylamide.html
29. http://grist.org/food/golden-rice-fools-gold-or-golden-opportunity/?utm_campaign=weekly&utm_medium=email&utm_source=newsletter&sub_email=blilliston@iatp.org
30. http://blogs.scientificamerican.com/guest-blog/2014/03/15/golden-rice-opponents-should-be-held-accountable-for-health-problems-linked-to-vitamain-a-deficiency/
31. www.organicconsumers.org/news/socialist-gmos
32. www.sciencemag.org/cgi/content/abstract/295/5555/674
33. www.nature.com/news/india-s-first-gm-food-crop-held-up-by-lawsuit-1.21303
34. www.theecologist.org/News/news_analysis/2140802/the_real_point_of_gm_food_is_corporate_control_of_farming.html;
35. www.upov.int/about/en/
36. http://tansey.org.uk/publications/control-of-food.html
37. www.foodandwaterwatch.org/insight/biotech-ambassadors
38. https://sites.google.com/site/gmodebate/trails
39. www.sciencedaily.com/releases/2016/11/161109181906.htm
40. www.oecd.org/agriculture/crp/42582878.pdf

41. http://news.bbc.co.uk/1/hi/special_report/1999/02/99/food_under_the_microscope/285408.stm& www.princeofwales.gov.uk/media/speeches/article-hrh-the-prince-of-wales-titled-questions-about-genetically-modified-organisms
42. www.reuters.com/article/2011/09/20/us-monsanto-superweeds-idUSTRE78J3TN20110920
43. www.monsanto.com/newsviews/Pages/india-pink-bollworm.aspx
44. www.ncbi.nlm.nih.gov/pmc/articles/PMC59819/
45. www.news.cornell.edu/releases/May99/Butterflies.bpf.html
46. www.theecologist.org/News/news_analysis/1745491/the_gm_lobby_and_its_seven_sins_against_science.html
47. https://sites.google.com/site/gmodebate/about-us/7-sins
48. www.tsl.ac.uk/news/independent-report-government-gm/
49. www.gov.uk/government/publications/genetic-modification-gm-technologies
50. www.tsl.ac.uk/news/independent-report-government-gm/
51. www.telegraph.co.uk/foodanddrink/7852762/Supermarkets-selling-meat-from-animals-fed-GM-crops.html
52. www.food.gov.uk/sites/default/files/public-attitudes-tracker-nov-14.pdf

CHAPTER 11

1. www.gov.uk/trade-tariff/commodities/0407210000
2. http://eur-lex.europa.eu/LexUriServ/LexUriServ.do?uri=OJ:L:2001:316:0005:0035:EN:PDF
3. http://pork.ahdb.org.uk/prices-stats/news/2017/february/a-recovery-for-bacon-sales/
4. http://meatinfo.co.uk/news/fullstory.php/aid/17978/Bacon_Report:_Healthy_drivers.html
5. www.bbc.co.uk/news/av/uk-politics-eu-referendum-36429170/brexit-and-the-great-british-fry-up
6. www.kelloggs.co.uk/en_GB/nutrition1/our-passion-for-nutrition.html
7. www.birminghammail.co.uk/news/showbiz-tv/two-sets-birmingham-twins-star-12418155
8. www.gov.uk/government/news/new-change4life-campaign-encourages-parents-to-be-food-smart
9. https://ec.europa.eu/unitedkingdom/news/uk-advertising-standards-authority-cautions-kelloggs-health-claims_en
10. http://sugarcane.org/global-policies/policies-in-the-european-union/eu-sugar-policy
11. http://farmsubsidy.openspending.org/GB/recipient/GB47951/tate-lyle-europe-031583/
12. https://blogs.spectator.co.uk/2017/02/brexit-mean-cheaper-food-dont-open-prosecco-yet/#
13. www.hydrocarbons-technology.com/projects/britishsugar/

14. https://cereals.ahdb.org.uk/markets/market-news/2016/october/18/prospects-2016-grain-market-outlook.aspx
15. https://cereals.ahdb.org.uk/media/1136831/2016-17-Early-Balance-Sheet-Final.pdf
16. www.theguardian.com/lifeandstyle/2013/oct/30/jam-wars-reducing-sugar-britain
17. http://eureferendum.com/blogview.aspx?blogno=84464
18. http://atlas.media.mit.edu/en/visualize/tree_map/hs92/import/show/all/0901/2014/
19. www.ico.org/documents/icc-107-7e-tariffs-trade.pdf
20. http://ec.europa.eu/taxation_customs/dds2/taric/taric_consultation.jsp
21. www.gov.uk/trade-tariff/commodities/1005900000
22. http://ec.europa.eu/taxation_customs/dds2/taric/measures.jsp?Lang=en&SimDate=20170302&Taric=0901000000&LangDescr=en
23. http://atlas.media.mit.edu/en/visualize/tree_map/hs92/export/show/all/0901/2014/
24. https://tinyurl.com/yak6hxhp
25. www.cityam.com/260597/sandwich-skills-shortage-pret-manger-boss-andrea-wareham
26. www.huffingtonpost.com/entry/brexit-pizza-chicago-ohare_us_576d4cc7e4b0dbb1bbba5048
27. https://tinyurl.com/glrbx3p
28. http://blog.policy.manchester.ac.uk/posts/2016/09/whats-in-your-brexit-burger-theres-even-less-chance-of-knowing-now/
29. https://sites.google.com/site/sustainablefoodreports/horse
30. www.fao.org/3/a-i4481e.pdf
31. www.gov.uk/trade-tariff/commodities/0902300000
32. https://tregothnan.co.uk/
33. www.telegraph.co.uk/news/2016/10/19/sneering-at-brexit-biscuits-shows-contempt-for-a-vital-british-i/
34. www.fginsight.com/news/news/school-milk-on-the-menu-after-brexit-16632
35. www.fginsight.com/news/news/school-milk-on-the-menu-after-brexit-16632
36. www.gov.uk/trade-tariff/commodities/0401201100
37. https://news.utexas.edu/2011/06/22/milk_studies
38. www.mirror.co.uk/news/uk-news/brexit-make-ice-cream-more-8135773
39. www.dailymail.co.uk/health/article-393432/The-chilling-truth-ice-cream.html
40. www.theguardian.com/world/2016/oct/19/eu-canada-free-trade-deal-delayed-belgium-ceta
41. http://researchbriefings.parliament.uk/ResearchBriefing/Summary/CBP-7492
42. www.gov.uk/government/publications/uk-food-and-drink-international-action-plan-2016-to-2020

43. www.forbes.com/sites/taranurin/2016/06/29/what-happens-to-beer-after-brexit/#f432d94330ee
44. http://goodbeerhunting.com/sightlines/2016/6/27/sightlines-uk-brewers-consider-the-implications-of-brexit
45. http://atlas.media.mit.edu/en/profile/hs92/1210/#Importers
46. http://atlas.media.mit.edu/en/profile/country/gbr/
47. www.scotch-whisky.org.uk/news-publications/news/brexit-what-now-for-scotch-whisky/#.WFlfwPmLRxA
48. www.decanter.com/wine-news/brexit-wine-prices-wsta-336341/
49. www.iwsc.net/result/trophy_winners/2016
50. www.independent.co.uk/topic/Champagne
51. www.independent.co.uk/topic/brexit
52. www.independent.co.uk/news/business/news/brexit-latest-news-british-champagne-france-sparkling-wine-leave-vote-food-drink-protection-rules-a7582981.html
53. www.cnbc.com/2017/01/27/american-beef-industry-sees-brexit-as-big-stakes-opportunity.html
54. www.fginsight.com/news/uk-wi-16639
55. https://ec.europa.eu/food/safety/chemical_safety/meat_hormones_en
56. https://fas.org/sgp/crs/row/R40449.pdf
57. http://beefandlamb.ahdb.org.uk/wp/wp-content/uploads/2016/07/UK-Yearbook-2016-Cattle-050716.pdf
58. http://beefandlamb.ahdb.org.uk/wp/wp-content/uploads/2016/07/UK-Yearbook-2016-Cattle-050716.pdf
59. https://ec.europa.eu/food/animals/welfare/practice/farm/pigs_en
60. www.npa-uk.org.uk/iqs/dldbitemid.135/dlsfa.view/Press_Releases.html
61. http://trade.ec.europa.eu/doclib/docs/2014/december/tradoc_152982.pdf
62. http://researchbriefings.parliament.uk/ResearchBriefing/Summary/CBP-7492
63. https://en.wikipedia.org/wiki/Ractopamine
64. www.breakingviews.com/features/cameron-china-and-the-missing-pig-semen/
65. http://pork.ahdb.org.uk/prices-stats/imports-exports/
66. www.gov.uk/government/publications/code-of-recommendations-for-the-welfare-of-livestock-pigs
67. www.ahdb.org.uk/brexit/documents/BeefandLamb_bitesize.pdf
68. www.sheepcentral.com/what-are-the-implications-in-uks-brexit-for-australian-red-meat-exports/
69. www.theguardian.com/politics/2016/sep/07/brexit-vote-great-relief-for-uk-fishing-industry-lords
70. http://ukandeu.ac.uk/british-fishermen-want-out-of-the-eu-heres-why/
71. *Telegraph*, 16 June 2016.
72. www.seafish.org/media/1653731/overview_-_brexit_and_the_uk_seafood_industry_1.3.pdf
73. www.un.org/depts/los/convention_agreements/texts/unclos/part5.htm

74. https://hansard.parliament.uk/lords/2017-01-16/debates/2705C651-928C-4266-9323-EC781E16F44D/BrexitFisheries(EUCReport)
75. www.gov.uk/government/news/environment-secretary-michael-gove-sets-out-new-approach-for-uk-fishing
76. www.seafish.org/research-economics/market-insight/market-summary
77. www.bbc.co.uk/news/magazine-17536764
78. www.irishtimes.com/opinion/brexiteers-solution-to-rising-food-prices-let-them-eat-prawns-1.2992272
79. http://scienceblogs.com/grrlscientist/2006/02/01/tsunamis-and-mangroves-the-shr/
80. http://neweconomics.org/2016/11/turning-back-to-the-sea/?_sf_s=blue+new+deal
81. www.food.gov.uk/sites/default/files/multimedia/pdfs/waterguideeng07updated.pdf
82. www.youtube.com/watch?v=Se12y9hSOM0
83. http://eur-lex.europa.eu/legal-content/EN/TXT/?uri=LEGISSUM:l21122b
84. www.fao.org/docrep/006/Y4343E/y4343e0i.htm
85. P.P. Courtney (1965), *Plantation Agriculture*, London, Bell's Advanced Economic Geographies, p.65.
86. www.pacificagricapital.com/

CHAPTER 12

1. B. Bragg (2006), *The Progressive Patriot*, London: Bantam.
2. https://en.wikipedia.org/wiki/Just_Transition
3. www.agrimoney.com/feature/uk-land-prices---how-will-they-fare-in-2017-amid-brexit-uncertainty--495.html
4. http://researchbriefings.parliament.uk/ResearchBriefing/Summary/CBP-7213
5. www.tandfonline.com/doi/pdf/10.1080/15693430601108086
6. www.nfuonline.com/news/brexit-news/eu-referendum-news/nfu-unveils-three-cornerstones-of-a-new-domestic-agricultural-policy/
7. http://europa.eu/rapid/press-release_IP-17-187_en.htm
8. https://ec.europa.eu/agriculture/survey
9. https://ec.europa.eu/agriculture/consultations/cap-modernising/2017_en
10. https://ec.europa.eu/agriculture/sites/agriculture/files/consultations/cap-modernising/summary-public-consul.pdf
11. www.pig-world.co.uk/news/new-agriculture-bill-in-queens-speech.html
12. www.eesc.europa.eu/?i=portal.en.press-releases.43610
13. J. Hanlon, A. Barrientos and D. Hulme (2010), *Just Give Money to the Poor*. Boulder, Co.: Lynne Rienner.
14. https://en.wikipedia.org/wiki/Agrarian_Justice
15. http://farmsubsidy.openspending.org/GB/
16. www.fwi.co.uk/news/small-farmers-demand-radical-shake-post-brexit-farm-policy.htm

17. www.sciencemag.org/news/2017/03/uk-scientists-prepare-impending-break-european-union
18. http://ec.europa.eu/trade/policy/countries-and-regions/countries/turkey/index_en.htm
19. www.confectioneryproduction.com/17771/news/bnf-survey-exposes-uk-students-ideas-food-healthy-eating/
20. https://sites.google.com/site/todmordenlearning/
21. https://sites.google.com/site/seeds4scotland/
22. http://farmandfoodmn.org/
23. http://sustainablefoodcities.org/
24. http://sustainablefoodcities.org/findacity
25. https://searchworks.stanford.edu/view/1874983
26. www.sussex.ac.uk/spru/newsandevents/2017/publications/food-brexit
27. www.telegraph.co.uk/news/2017/03/14/buy-british-button-could-let-online-shoppers-filter-foreign/
28. www.countrysideonline.co.uk/home/
29. www.gov.uk/guidance/public-sector-procurement-policy
30. www.gov.uk/government/publications/a-plan-for-public-procurement-food-and-catering
31. www.nfuonline.com/news/featured-article/back-british-farming-brexit-and-beyond-the-nfu-20/#top
32. www.theguardian.com/commentisfree/2017/jun/26/brexit-watershed-farming-food-industry-michael-gove
33. www.choicesmagazine.org/2003-4/2003-4-01.htm

Index